How The Earth Was Formed
Third Edition

Max B. Frederick
An Old Scientist

"And the Earth was without form and void. And Darkness was upon the face of the deep..."
The second verse of the bible Genesis 1:2

"Ideas without precedent are generally looked upon with disfavor and men are shocked if their conceptions of an orderly world are challenged."
J. Harlen Bretz, 1928

Cover Photo: Many times the creation accounts of the bible mention the first light. In the book of Job the first event that resulted in that light is described as a shout, in 1950 Fred Hoyle ridiculed the idea of a beginning being such an explosive event by calling it the "Big Bang." Whatever it was it resulted in the first generation stars, (morning stars) also mentioned by Job. In 2006, scientists at Goddard Space Flight Center actually photographed the glow from those first generation stars produced by that "Shout" or "Big Bang." The universe is wrapped in it. That actual photograph is what you see on the cover of this book. Courtesy NASA/JPL-Caltech, as of the writing of this book, that image may be viewed on the Internet at, http://www.spitzer.caltech.edu/images/1697-ssc2006-22a-The-Universe-s-First-Fireworks

Published in the United States of America.

First Edition, November 2014
Second Edition, September 2015
Third Edition, May 2017

ISBN: 978-1546647423

How The Earth Was Formed

Max B. Frederick, Publishing,
146 Laurel St.
Central Point, Oregon 97502,
United States of America

Ordering information to be found on the web at:
www.EyewitnessToTheOrigins.com

How The Earth Was Formed
Third Edition

Discovering What the Bible Really Says About the Origins and Comparing it to Reality as Discovered by Modern Science

Max B. Frederick
An Old Scientist

"And the Earth was without form and void. And Darkness was upon the face of the deep..."

The second verse of the bible Genesis 1:2

Max B. Frederick, Publishing,
Central Point, Oregon,
United States of America

Third Edition May, 2017

If the bible is credible in areas related to science that can be verified, What, then do you do with what the bible says about eternal life?

Table of Contents

The author recommends that the book, *"The Mystery of Psalm 104"* be read before this book in preparation for many of the concepts talked about in this book. *"The Mystery of Psalm 104"* may be purchased on the Internet by following the links from www.EyewitnessToTheOrigins.com

Preface

> *"And the Earth was without form and void. And Darkness was upon the face of the deep..."*

The second verse of the bible, Genesis 1:2 (KJV)

The inspiration for the name of this book comes from the second verse in the bible.

That verse refers to a specific time, place, and situation in the history of the formation of the world we live in. It refers to a specific event in the history of our planet. That verse was placed up front in the bible to be an introduction to a series of accounts that tell the history of how the earth was formed.

Somehow, down through the ages, the original meaning of that verse became lost. Then people thought they found it again. From science they learned that at one time in the early history of our planet, it was a chaotic mass of swirling gasses. Formless they thought. Wow, the bible is right. Science confirms it. But now we know that is not what that verse is talking about at all. It is much more specific.

This book compares knowledge of pre-historic things from two dissimilar sources. It presents knowledge from both sources in chronological order and shows the remarkable similarity demonstrating the fact the ancient source of knowledge had it right long before modern science discovered the same things. The first source is the written record in ancient scriptures of the bible. The second source is the evidence that is recorded in geology and astronomy. This information is presented here as a timeline of how everything around us came to be what it is today.

Before the beginning of the twentieth century, around the year 1900, nobody knew that verse was an introduction to several biblical accounts that told the same story as this book tells.

Scientists back then had no knowledge of the specific event it was referring to. That event, and the circumstances around it, had not yet been discovered by modern science and the long ago understanding of what it meant had long since been lost and the verse was simply preserved and carried along as some religious theobabble by dedicated theologians.

How the Earth was Formed

Theologians did not recognize it for what it said. They did not even imagine such an event was in the history of our planet. They assumed it referred to something else. This and many other biblical references referring to the same thing appeared to talk of water covering the earth. Theologians assumed they were references to the flood of Noah. They did not even suspect they referred to an earlier more significant event that is now becoming realized by modern science.

Scientists did not understand the significance of the event that verse referred to. Little did they know that the continents are temporary in the sense of the long run. Little did they know that the continents are floating, buoyed up by the mantle below. Little did they know that there had been a time when the entire surface of this planet had been covered with water before the continents were formed. It is the dry land of those continents that verse refers to, not the planet earth.

It should be obvious. The bible itself defines the word *earth,* to be the dry land. It is described in detail in the ancient scriptures:

And God said, Let the waters under the heaven be gathered together unto one place, and let the dry [land] appear: and it was so. And God called the dry [land] Earth; and the gathering together of the waters called he Seas: and God saw that [it was] good. Gen 1:9,10

You see, the word "Earth" seen in the second verse of the bible, does not refer to the planet, but to the solid surface parts of the planet as opposed to the oceans. It refers to the "dry land" of the continents. At one time in the history of our planet, the "earth" was not formed into the dry land of the continents. And now it is. Currently, only about 29 percent of the surface of out planet is above sea level. The continents are isolated piles of earth piled up over three miles high above the sea floor. That leaves the rest of the solid surface of the planet—about 71 percent of it, to be a sea floor an average of over two and a half miles below sea level.

It was a big deal, and still is. As of yet, the mechanism that caused the "forming of the earth,"—the gathering of the dry land into continents,— is still little understood. That mechanism still operates today, keeping the dry land piled up into continents so they do not spread out over the sea floor under the ocean. It is a situation that is unique to our planet alone among other nearby heavenly bodies studied by scientists.

6

Introduction:

How the Earth was Formed
By Max B. Frederick, AnOldScientist

"Ideas without precedent are generally looked upon with disfavor and men are shocked if their conceptions of an orderly world are challenged."

J. Harlen Bretz, 1928

This writing is about the events and processes the universe went through in the formation of the earth, sea, sky, and ecology around us. It is about the information recorded in the bible concerning these processes. It is also about the inability of theologians to see it, and how to dispel their misinterpretation of what they do see, both in their holy writings and in the world around them. It is driven by how this affects the choice of intelligent humans to believe the bible or not, and the consequences of that choice.

Two Credible, Independent Sources of Information:

There appears to be two credible sources of information about how the earth was formed. Those two credible sources are very dissimilar in nature. The first is highly ridiculed by keepers of the second for being so out of touch with reality. The first was recorded long ago, but as the ages passed, the understanding of it was tainted by the lack of technical understanding by the keepers of that information. So much so was it misinterpreted that it lost all credibility among critical scholars. Thus, the independence of the two sources. Yet, careful analysis indicates they are in full agreement as to technical details.

Ancient Writings:

The first credible source of information is a specific collection of ancient writings. Some of these writings appear to be as ancient as writing itself. These ancient writings contain descriptions of, and references to, things that are beyond the possibility of having been observed by human beings because they happened long before human beings existed. The fact that this information was recorded long before modern science existed to discover the same information confounds those who would like to believe that no such credible source could exist. Those ancient writings themselves claim their ultimate source is a communication from some extra-terrestrial intelligence that was instrumental in causing the earth to form.

This claim cannot be tested by the scientific method. But the credibility of the information in that claimed communication can be so tested and is verified by modern science, and lends credibility to this claim that cannot be so tested.

Modern Science:

The second credible source is in the realm of modern science. It is observation of physical reality as processed through the scientific method. In this writing it shall be referred to simply as, "modern science."

In the realm of modern science there are two physical sources of information.

The first physical source of information in the realm of science is evidence from the earth itself. The record of its formation is recorded in the rocks of the earth. This record is called the geologic column.

The second physical source of information in the realm of science is evidence observed in the cosmos. That information is contained in the light currently reaching us from far away objects in the universe. Due to the speed of light, that information originated in the ancient past and is arriving here now where we can currently see processes that happened in the formation of other heavenly bodies.

Exploiting the Information:

In this book it is the first source of information that is explored and compared to what modern science has discovered in recent times from the second, physical source. The exploitation starts with finding all the biblical accounts of the beginnings—and there are about three dozen major accounts of it—and interleaving them into one long single chronology using clues internal to those biblical accounts.

The fact that this written information exists is evidence of a credible ultimate source from which it came. The fact that this information was preserved in religious writings through times when much of this credible information was considered to be incredible—and ignored— or misinterpreted—is astounding. It was simply preserved for us at

---

this time because the writings were considered to be sacred religious writings.

The Sources Converge:

As new information is being discovered by modern science, and as old information is being recognized by bible scholars, the information from both modern science and the ancient scriptures is converging on reality. This convergence is interesting to watch. In the past, the information recorded in the ancient writings has been obscured by religious interpretation. Only recently, with the light of significant discovery by modern science, has the reality begun to shine through the religious misunderstanding. As long as we avoid religious interpretation of either the information in the ancient scriptures or of the information discovered by modern science, the information from both sources is converging on the same reality

Organization of this Book:

There have been various stages of development in the long process of the formation of the earth, sky, sea, and ecology we see in the world around us. In this book, the stages and significant details are set in the context of the timeline of all time from eternity past to eternity future. Here they are presented in the order of their occurrence. At the beginning of each section, a portion of the complete list is presented. The complete list is presented in Appendix A, along with explanation of the numbering system.

According to modern science, there are things that are considered significant that are not included in the list of items found in the bible. One such thing is the catastrophic collision of another heavenly body with the planet earth resulting in much rock material being torn off the planet earth and becoming the moon. Another is the long period of time, early in the planet's development where the water that covered the surface of this planet was all ice, similar to Europa, a moon of Saturn.

According to the bible, there are other things that are significant that are not included in the list of items modern science considers to be important. Such things include the claim that it was all the result of the creative work of some intelligent, extra-terrestrial life form.

And then there are many items common to both the biblical accounts and the knowledge base of modern science. Many are items that modern science did not have on their reality list until recently, until within the past three hundred fifty years. In the mean time, modern science has independently discovered these items and have added them to their own list, not knowing the bible published them thousands of years earlier. An example of this is the statement in the bible that the mountains go up and the valleys go down[1]. Discovered around 1900, that is now called the principle of isostacy by modern science. But before that was discovered theologians could not bring themselves to translate it correctly, "knowing" that was not true. Instead they fudged the translation to imply it was the water that was going up by the mountains and down by the valleys[2].

And then there may be more details in the biblical accounts that modern science has not yet discovered, and we do not yet recognize. That is the track record of the past. Theologians had information at their fingertips and did not recognize it, instead, misinterpreting it to be some meaningless theobabble. Some of that discovery has happened since the author started writing on the subject of science and the bible some twenty years ago.

This book is not about evolution. Remember, even though evolution is currently (temporarily) a major tenant of modern science, modern science has not discovered life arising from spontaneous generation. The closest they can get to it is to have an intelligent being (scientist) set up conditions to cause what appears to be spontaneous generation, but in reality only proves that an intelligent being can cause what appears to be spontaneous generation. The bible is emphatic that humans were the product of the creativity of an intelligent being. Modern science has not discovered anything to change that.

Why is it Important?

Science or religion? Which is a friend of the bible? Which is a friend of reality? That is a significant question. Why is that so? Because it is not Science versus the Bible. What it boils down to is

---

[1] Psalm 104:8 (ASV) Published 1901 *"The mountains rose, the valleys sank down..."*
[2] Psalm 104:8 (KJV) Published 1611, 1769 *"They go up by the mountains; they go down by the valleys..."*

the reality of science and the bible on one side of the debate, versus the fiction of traditional interpretation by religion on the other.

In the realm of what modern science can actually verify, there is very little difference between what the bible has to say and what modern science has discovered to be fact.

That last sentence was worded carefully. Read it again. It does not say anything about the opinions expressed by either theologians or scientists. The animosity and disagreement is not between the facts discovered by modern science and the facts as recorded in the ancient scriptures of the bible—it is between the opinions of theologians and the opinions of scientists.

There is one major difference—the intelligence, or lack of intelligence of the ultimate cause. The ancient scriptures record as fact the concept that it was all created by some plural form of intelligent energy, force, power, etc. called *Elohiym* in the Hebrew language. The accepted opinion of Modern science is in accord with the idea of some form of energy, force, power, etc. existing in the beginning, and causing the beginning of the universe, but the generally accepted interpretation of modern science rejects the idea of that cause having intelligence.

So the question remains. Is there a living intelligence, a living form of intelligent energy that is the creator who has communicated to humans the information recorded in the ancient scriptures as to facts, processes, and events, particularly the chronology involved in the creation of the world we see around us?

The evidence is in the fact that such information exists in the holy writings of the ancient Hebrews—information that could not have gotten there from the discoveries of a modern science that did not exist until thousands of years later. That evidence indicates the existence of some form of living extra-terrestrial intelligence was involved.

Organization of Detail

A table summarizing significant items common to the bible and modern science is found in Appendix B.

The biblical outline of all existence is expressed somewhat differently than typical of modern science, but is in strict accord with the same time sequence of events. The biblical outline consists of five major eons:

Eon of Eternity Past – Timeless Past.

Eon of Early Development – Begins with the biblical, "insemination of the cosmos" (the modern science, "Big Bang".)

Eon of Preparation for Complex Life /Ecology – Begins with the emergence of the continents. (Roughly equivalent to the beginning of the actual geologic column.)

Eon of Complex Life Forms – Begins with Cambrian Explosion, Ends with the future biblical, "consummation of the Eon." (the modern science undefined end of the universe.)

Eon of Eternity Future – Timeless Future, including "The Judgment"

Within that biblical outline, within the ancient writings, there are found references to over a hundred details arranged in the same chronological order as has been discovered by modern science.

A detailed table of biblical items, along with an explanation of the numbering system used in this writing, is found in Appendix A.

Suffice it to say at this point that in the numbering system the parts of the number before the double decimal are related to chronology and the parts after the double decimal are not necessarily in any chronological order, but are within the same chronological period.

Eon 1: Eon of Eternity Past-Prior Existence

Brief outline of Eon 1:

1.0. Eon of Eternity Past

1.0.0   Eternity Past Existed

1.1.0   Conditions Were Set Up to Start Universe

The Eon of Eternity Past is first of five Eons in the chronicle of existence from eternity past to eternity future. This is the eon before the event that started the current universe.

There is a lot of speculation as to what existed, and what happened before the current universe started.

In the realm of science, such speculation is a relatively new development. Prior to the 1950s, most scientists believed in the steady state theory of the existence of the universe. That is, they believed the universe always existed kind of like it exists now. Sometime in the 1960s the general attitude changed toward the realization that the universe had a beginning at a specific place at a specific time in the past. After the realization that there actually was a beginning became widespread, speculation about what it was like before the beginning has been a lively subject.

In the realm of the bible, it has always been evident that the bible claims that God existed before the universe, and that God will continue to exist after the universe has expired. But the bible has much more to say about the physical nature of the pre-existing "God" than most theologians realize.

Outline[3] of Eon 1:
1.0. Eon of Eternity Past
1.0.0   Eternity Past Existed
1.0.0..1 Eternity Past Existed
1.0.0..2 Before the Universe existed
1.0.0..3 Something existed in the absence of the universe
1.1.0   Conditions Were Set Up to Start Universe
1.1.0..1 That Something caused the Universe to come into existence
1.1.0..1.1 Pre-determined by intelligence(Logic, Laws of Physics)
1.1.0..1.2 Specific pre-existing conditions reacted to Laws of Physics
1.1.0..1.3 Universe Constructed (created) by Energy, Power Force

The outline above has eight significant details about existence in the Eon of Eternity Past. These details are worded in a way that should be somewhat acceptable to both scientists and to theologians. Complete acceptance is not the goal here, only a common ground of understanding of what each other is talking about.

The Science

The eight significant details above are presented below with slightly different wording:
1. Eternity Past Existed.
2. It existed prior to the "beginning" (Big Bang?/Creation) event.
3. Something, not nothingness, existed in the absence of the universe.
4. Conditions were set up to start the universe.
5. The Something that existed caused the Universe to come into existence.
6. Universe was pre-determined by Logic, Laws of Physics, intelligence.
7. Specific pre-existing conditions reacted to Plan/Laws of Physics.
8. Universe Constructed/created by/from some Energy, Power, Force.

Though some may say that each and every one of the points of detail in The Eon of Eternity Past are outside the realm of modern science, modern scientists do have something to say concerning all of them.

Concerning the first two points, eternity past did exist, and it did exist before an event commonly accepted as the beginning of the current universe, modern science is rapidly coming to the conclusion that this Eon of Eternity Past actually did exist. It existed before the

---

[3] *Complete outline is found in Appendix A: Detailed Chronology List*

cosmos, before there were any oceans or continents before any physical existence, of human life.

Concerning the third point, just what it was that did exist, science has varied opinions on this matter.

A typical Modern Science saying states that the universe is the interdependent existence of time, matter, space and energy. Before the universe came into being, during the eternity past eon, some scientists say that time, as we know it, may or may not have existed, perhaps in a different dimension.

But modern scientists seem to be coming to a consensus that there really was a something[4], and that some something really did happen, and it really did set up the conditions that caused the event that is popularly known as the beginning of the universe.

Just what this something was is still a matter of debate.

A few years ago, when scientists were beginning to realize that the current universe really had a beginning, some scientists still believed it had existed forever. That was known as the steady state theory of the universe's existence. Others began to believe it had a beginning. In the 1960s the event that was becoming accepted as the beginning of the universe became commonly known as the "Big Bang," a term of derision[5] ridiculing those scientists who believed in such a preposterous thing.

In more recent times that event known as the Big Bang, has come to be commonly accepted by modern science as an early event in the formation of this universe, but not the absolute beginning—there had to be something that set up the conditions for the event that started the current universe.

Current speculation is in the realm of what existed, and what it was that set up the conditions for that event to occur. Some say that there was a black hole that exploded and continued to expand to become

---

[4] *See Appendix C: Does Nothingness Exist?*

[5] *In about 1950 The term, "Big Bang" was coined by astrophysicist Fred Hoyle as a term of derision because he was championing the competing theory of a Steady State Universe. It was not until after 1964, after the discovery of cosmic background radiation, that the competing theory of a Steady State Universe was overwhelmed by the theory that the universe actually had a beginning.*

the universe we see today. But, what caused that black hole, and what caused it to explode is still the subject of developing theories[6].

Currently, it appears that it is the considered opinion of respected scientists that before the event of the beginning of the universe, conditions were set up to cause (or to allow) that event to happen.

That opinion includes the idea that there had to have been existence of some unknown nature before the beginning of the universe. That unknown existence had to have had the capability to set up the conditions for that event to occur.

Whoever or whatever set up those conditions gave it all the order necessary to develop into the universe of today.

At least in the first five details:—when simply stated this way—all three sources of information are in full agreement.

A difference of opinion

On the next three points, and these three only, there is a difference of opinion.   Interpreters of the cosmos and geologic column (scientists) would state them one way, and interpreters of the ancient writings (theologians) would state them another. They all agree that all three points exist, they would just state them a different way.

A scientist might state them in a way that implies things just spontaneously happened.

A theologian might state them as being caused by some extra-terrestrial form of life. Perhaps it was some self-aware form of living energies with the additional aspects of intelligence, and wisdom (engineering.) Ancient writings in the original language describe it as such, and call it the *Yehovah Elohiym*[7] (literally meaning, eternally existing energies). For simplicity here I will simply refer to this existence as "The God of the bible", or "God" for short.

_____

[6] *Developing theories described in* Scientific American, *"Following the Bouncing Universe," by Martin Bojowald, October, 2008, p. 44.*

[7] *Theologians have degraded the meaning of this existence to a religious "LORD God", a name devoid of all original physical meaning and only referring to status as a boss in charge who has been elevated from a lesser status. See Glossary See Glossary on the internet at,   www.EyewitnessToTheOrigins.com/glossary for fuller explanation of the reality of meanings of biblical terms that have been religiousized and thereby had their meanings obscured.*

Modern science has expressed the opinion that the order we see in the universe of today is the result of spontaneous generation.

The bible states emphatically that the order necessary for that existence to develop into the universe of today is due to the planning and designing, and execution by God.

The bible is emphatic in the three attributes of the prior existence. The ancient scripture says these attributes of the pre-existence of God, are, (1) the Planner (Intelligence,) (2) the Designer (Wisdom) and (3) the Creator (Powers), as stated in these last three points of detail.

More agreeable to the mind of the scientist, these three points of detail could be replaced with the speculations:

1.1.0..1.1   The Laws of Physics,—Logic/intelligence— that caused the universe to come into existence was one of the attributes of the unknown existence before the beginning of the universe.

1.1.0..1.2   That setting up of the conditions that caused the universe to come into existence— The prudent application of those Laws of Physics; the design/wisdom that engineered the universe into existence—was one of the attributes of the unknown existence before the beginning of the universe, even if it was all spontaneously by accident.

1.1.0..1.3   Some extreme force/energy/power that caused the universe to come into existence was one of the attributes of the unknown existence before the beginning of the universe.

The Ancient scripture version of these three points of detail is:

1.1.0..1.1   Planned by intelligence [Scientist aspect of pre-existence: Laws of Physics, Logic]

1.1.0..1.2   Designed by Wisdom [Engineer aspect of pre-existence]

1.1.0..1.3   Constructed (created) by Power [Authority (ruler, king, lord, father) aspect of pre-existence]

A compromise version might look something like this:

1.1.0..1.1   Pre-determined by intelligence (Logic, Laws of Physics)

1.1.0..1.2   Specific pre-existing conditions reacted to Laws of Physics

1.1.0..1.3   Universe Constructed (created) by Energy, Power Force

---

A universal claim in all the creation accounts of the bible is that God did it.

In reality, the only difference between the opinion of modern science and the detail presented in the bible is the bible presents it as an intelligent designer and names it the EXISTENCE (God.)

Modern science leaves it open to the possibility of spontaneous generation, but rejects the idea of an intelligent designer,

However, any claim or even hypothesis as to whether it was God, or it was spontaneous generation, is outside the realm of discovery by the scientific method.

Even though the claim that God did it is outside the realm of science, this claim is completely within the realm of the bible.

The realm of the bible is the realm of an eyewitness. An eyewitness establishes his credibility by consistency in truth where it can be verified by science, and then his realm of credibility is extended to what has been observed first hand.

The God of the bible claims to have been there and witnessed it. Modern science makes no such claim.

The irony is that any experiment to prove spontaneous generation is also outside the realm of science.

Application of the scientific method to any experiment concerning spontaneous generation being responsible for the beginning is beyond the scope of science. In the far out chance that scientists could set up a condition where there were no universe, and then set up the conditions that would cause such an event as the beginning of the universe, it would only demonstrate that an intelligent being could cause the beginning of the universe. Therefore, such an experiment to prove spontaneous generation is beyond the scope of science. But to prove the *possibility* of intelligent design is not outside the scope of science.

The bible claims eyewitness knowledge of creation.

There is no one who claims eyewitness to spontaneous generation.

But we all agree. Something did happen. We exist. So then, lets accept the fact that the bible and modern science agree in what happened, even though they may disagree as to what or who did it.

The Bible

In reading the following passages of scripture, concentrate on the remarkable agreement in detail as to what happened and realize this remarkable agreement actually exists even though it was recorded in the ancient scriptures long before the fact of that same information was actually discovered by modern science in recent times. The writers of the ancient scriptures had no earthly way of knowing the facts they say they got from "God."

> *But the LORD [is] the true God, he [is] the living God, and an everlasting king: at his wrath the earth shall tremble, and the nations shall not be able to abide his indignation. Thus shall ye say unto them, The gods that have not made the heavens and the earth, [even] they shall perish from the earth, and from under these heavens. He **hath made the earth by his power**, he hath **established the world by his wisdom**, and hath **stretched out the heavens by his intelligence**.*

Jeremiah, c. 585-580 BC.,
The True God, Triune Creator Account,
A Biblical Creation Account: Jeremiah 10:10,12

Note: This same account is repeated in: Jeremiah 51:15[8]

The opening line of the third creation account of Moses further illustrates that same triune nature of that pre-existence that caused the order of the universe:

> *"**And God said**,*   [Planning (Intelligence)]
> ***Let there be** light:* [Engineering (Wisdom)]
> *and there was light."* [Constructing (Power)]

Moses, ca. 1445 BC., from
A Creation Account Illustrating the Fourth Commandment,
A Biblical Creation Account: Genesis 1:3   {KJV)

> *Before the mountains were brought forth, or ever thou hadst formed the earth and the world, even from Everlasting-to-Everlasting, thou [art] God.*

---

[8] *He hath made the earth by his power, he hath established the world by his wisdom, and hath stretched out the heaven by his understanding. (Jer 51:15 KJV)*

Moses, ca. 1445 BC., *The Everlasting-to-Everlasting Account*
Moses Creation Account 1: A Biblical Creation Account: Psa
90:2 {KJV)

²⁵ᵃ *In ages past*
  ²⁵ᵇ *thou hast laid the foundation of the continents:*
  ²⁵ᶜ *and the heavens [are] the work of thy hands.*
David, c. 1015 BC, *Ages Past and Future*,
A Biblical Creation Account: Psalms 102:25

²² *The LORD possessed me in the beginning of his way, before*
  his works of old.
²³ *I was set up from everlasting,*
  from the beginning,
  or ever the earth was.
Wisdom, c. 970 – 930 BC,
Eyewitness Account by Wisdom,
A Biblical Creation Account: Proverbs 8:22,23

¹*In the beginning was the Intelligence, and the Intelligence was*
  with God, and the Intelligence was God.
²*The same was in the beginning with God.*
³*All things were made by him; and without him was not any thing*
  made that was made.
John, c. 80-95 AD, *Beginning of Time*,
A Biblical Creation Account: John 1

¹:⁸ *I am Alpha and Omega, the beginning and the ending, saith*
  the Lord, which is, and which was, and which is to come, the
  Almighty.
¹:¹¹ᵃ *Saying, I am Alpha and Omega, the first and the last...*
⁴:⁹ *...to him that sat on the throne, who liveth for ever and ever,*
⁴:¹⁰ *The four and twenty elders fall down before him that sat on*
  the throne, and worship him that liveth for ever and ever, and
  cast their crowns before the throne, saying,
⁴:¹¹ *Thou art worthy, O Lord, to receive glory and honour and*
  power: for thou hast created all things, and for thy pleasure
  they are and were created.

5:14b ...And the four [and] twenty elders fell down and worshipped him that liveth for ever and ever..

10:6 And sware by him that liveth for ever and ever, who created heaven, and the things that therein are, and the earth, and the things that therein are, and the sea, and the things which are therein...

15:7b ...God, who liveth for ever and ever.

21:6 And he said unto me, It is done. I am Alpha and Omega, the beginning and the end. I will give unto him that is athirst of the fountain of the water of life freely.

21:7 He that overcometh shall inherit all things; and I will be his God, and he shall be my son.

22:13 I am Alpha and Omega, the beginning and the end, the first and the last.

John, c. 90-96 AD, *Time and Eternity,*
A Biblical Creation Account: Revelation

That ancient scripture says within this Eon there existed the one true God. The same ancient scripture refers to this one true God by a name that means the EXISTENCE.

Not much is said about what happened in this Eon other than it is endless and the true God is an endless EXISTENCE therein with the attributes of Intelligence, Wisdom, and Power.

"Have you never known? Have you never heard? The God who exists without ends of time, the EXISTENCE, who created the earth with ends [of time], who does not grow weak, who does not wear out, His intelligence is without limit".[9]

Isaiah, c. 700-680 BC, *An Account of Early Planning and Detail,*
A Biblical Creation Account: Isaiah 40:28

*"But the EXISTENCE [is] the true God, he [is] the **living God, and an everlasting** king: at his wrath the earth shall tremble, and the nations shall not be able to abide his indignation. Thus*

---

[9] *King James Version:* Hast thou not known? hast thou not heard, [that] the everlasting God, the LORD, the Creator of the ends of the earth, fainteth not, neither is weary? [there is] no searching of his understanding. *(Isa 40:28 KJV)*

*shall ye say unto them, The gods that have not made the heavens and the earth, [even] they shall perish from the earth, and from under these heavens. He hath made the earth **by his power**, he hath established the world **by his wisdom**, and hath stretched out the heavens **by his intelligence.**"*

Jeremiah, c. 585-580 BC.,
The True God, Triune Creator Account,
A Biblical Creation Account: Jeremiah 10:10,12

It is plain to see that the bible says that within this Eon, something EXISTED that set up the conditions for the universe to come into existence.

The universe was planned by the *Intelligence* of that EXISTENCE.

The universe was engineered by *Wisdom* of that EXISTENCE.

And the universe was brought into existence by the *Power* of that EXISTENCE.

Summary of Detail:

Within this Eon chronology is a strained concept. Time as we know it may not have existed in what some believe to be the timeless past of the Eternity Past Eon. Yet, chronological significance may be assigned to the concept that condition were set up to start the universe before the universe started. How long before is a wasted question in the absence of time as measured by physical motion of the universe. It may have been infinitely long—it may have been infinitely short.

The points of detail assigned to the timeless past of the Eternity Past Eon include: [10]

---

[10] A physical existence of a pre-universe physical life form is not a point of detail listed by The Old Scientist.

Concerning the pre-existence of a physical life form of some kind before the beginning of the universe, that is outside the realm of testing by scientific method. It is a claim of the bible alone, if in fact that is what the bible claims. It may be a matter of opinion as to whether the bible makes this claim. That claim may only be implied from the fact that in Psalms 148:2 the bible mentions the angels and his hosts before listing the components of the universe in an ordered list of over two dozen items which are accurately listed in a chronological sequence. One may argue the angels are part of that ordered list, or they are a

     23

1.0. Eon of Eternity Past

1.0.0 Eternity Past Existed

1.0.0..1 Timeless [endless] Eternity Past Existed

1.0.0..2 Before the Universe existed

1.0.0..3 Something (God) existed in the absence of the universe [Intelligence, Wisdom, and Power]

1.1.0 Conditions Were Set Up to Start Universe

1.1.0..1 That Something (God) caused the Universe [Everything] to come into existence

1.1.0..1.1 Planned by intelligence [Scientist aspect of pre-existence: Laws of Physics, Logic]

1.1.0..1.2 Designed by Wisdom [Engineer aspect of pre-existence]

1.1.0..1.3 Constructed (created) by Power [Authority (ruler, king, lord, father) aspect of pre-existence]

Or alternatively: (According to Modern Science)

1.0. Eon of Eternity Past

1.0.0 Eternity Past Existed

1.0.0..1 There was a timeless or endless (everlasting) Eternity Past.

1.0.0..2 It existed before the universe came into existence.

1.0.0..3 Something unknown existed in the absence of the Universe.

1.1.0 Conditions Were Set Up to Start Universe

1.1.0..1 That unknown something caused it—set up the conditions for the universe to come into existence.

1.1.0..1.1 The Laws of Physics—The Logic/intelligence that caused the universe to come into existence was one of the attributes of the unknown existence before the beginning of the universe.

1.1.0..1.2 The prudent application of those Laws of Physics—That setting up of the conditions that caused the universe to come into existence; the design/wisdom that engineered the universe into existence—was one of the attributes of the unknown existence before the beginning of the universe.

1.1.0..1.3 The power that made it happen—that caused the universe to come into existence was one of the attributes of the unknown existence before the beginning of the universe.

---

preamble to that list. Or one may argue that angels are not physical, rather are extra-terrestrial forms of life. A factor in the argument may be that angels are messengers and absent someone to send a message to there would be no messenger.

　　　　　24

Science rightly claims no human (nor scientist) was there to be an eyewitness. The bible claims to be the record of someone (not human) who was there and therefore, has the status of eyewitness.

Therefore, the question is: Can, and does the validity of the information presented by the bible substantiate its credibility as coming from an eyewitness? Apparently it is so.

Eon 2: Eon of Early Development

This Eon stretches over a vast length of time and contains six significant phases of the development from the beginning of the universe to when the planet earth was completely covered with waters of the ocean shortly before the continents finally emerged to be and to remain above sea level.

The following brief outline of this era lists the six phases and events that occurred within those phases of this eon.

2.0. Eon of Early Development

2.1. The Universe Formed - *The Heavens*

2.1.0.. The beginning of Universe was a specific event, the beginning of time, space, matter and energy.

2.2. The Expansion of the Universe

2.2.1 Then there was the expansion of the universe.

2.2.2.. Water became abundant in outer space

2.2.3.. The stars came into existence

2.2.4.. The Sun came into existence

2.3. The Development of the Solar System - *The Chambers of the South*

2.3.1. Then the solar system developed

2.3.2. Then hydrologic cycle (water cycle) developed in outer space of the solar system.

2.4. The Development of the Planet Earth

2.4.1.. Then planet earth developed with molten magma surface (mantle.)

2.4.2.. Then after a while the earth cooled enough to develop surface water

2.5. The Development of the Atmosphere

2.5.1.. Then an atmosphere developed.

2.6. The Development of the Oceans - *The Deep*

2.6.1.. Then the atmosphere captured (separated, set apart) water from outer space.

2.6.2.. Then the earth was covered by the oceans,

This second Eon begins at the point in time when the universe came into existence in the great event that is sometime referred to as the "Big Bang." And it ends when the planet earth was covered with the waters of the ocean, before the continents emerged to be above sea level. During this eon there seems to be an emphasis on the water, a fact that modern science is only currently realizing to be true.

Both of those events, the beginning of the universe, and the beginning of the continents, appear to be very significant.

There may be other significant events discovered by modern science that occurred during this eon that could be added to this list. But there appears to be no mention of them in the bible, they are not listed. If found, they may be added later. This book is not infallible.

During the last several years, scientists have learned a lot about the Early Development of our universe. Much of it was not known before the great separation of science and religion—about 375 years ago. Even more so during the last half of that time—since the favorite weapon[11] of those bent on proving there is no God was first published.

Ironically, much of what scientists have discovered about the Early Development of our universe, during those few short years, the bible had already published.

According to modern science, there are events that occurred within that Eon that I have not yet found a reference to, such as the glancing blow collision of another large heavenly body with the planet earth, stripping off much of the outer layer(s) of less dense material and causing the formation of the moon. Yet the gathering into the continents of what lighter material that is left on the surface is one of the big deals related in multiple biblical creation accounts. Ironically, it is this event that makes the gathering of the continents possible.

The Eon of Early Development encompasses all time from the creation of the universe through the filling of the oceans. Within this eon the bible mentions over fifty important points of detail of interest to science. Amazingly, (amazing only if you do not believe in something special about the bible) the biblical accounts place them all in the exact chronological order that modern science has discovered. And all these details occurred before there were even any continents.

All of this occurred and was recorded in the ancient scriptures of the bible. It was a long and eventful era and recorded in great detail. Yet, most theologians have no idea that it is even mentioned in the bible.

---

[11] *The theory of evolution was first published in 1859, only about one hundred fifty years ago. Darwin, C., 1859.* The Origin of Species by means of Natural Selection or the Preservation of Favoured Races in the Struggle for Life. *London, John Murray, first edition.*

27

2.1 The Beginning of the Universe

The beginning of the universe is the first phase in Eon 2, the Eon of Early Development.

The universe had a beginning. The event of that beginning was first published in the bible, along with several points of detail. All this was later discovered by modern science. It was a single event with several points of detail as outlined below.

Outline[12] of Eon 2.1:

2.1.   The Universe Formed - *The Heavens*

2.1.0..   The Forming of the Universe was a specific event,
It was the beginning of the current universe of interdependent time, space, matter and energy.

2.1.0..1.   The Universe had a beginning. It has not existed forever.

2.1.0..2.   It happened a long time ago

2.1.0..3.   Light was one of the first things to come into existence (Light existed from the "beginning")

2.1.0..3.1. Light had a beginning. It has not shone forever.

2.1.0..3.2. Light had a beginning from "darkness"

2.1.0..4.   It was an explosive event.

2.1.0..4.1   The bible describes it as a great blowout.

2.1.0..4.2   Modern science refers to it as a "Big Bang."

We exist.

The universe exists.

The bible says we exist and the universe exists.

Science says we exist and the universe exists.

There may be a few far out philosophers that question our existence. But no one seriously disputes our existence.

But how long has existence (the universe) existed? That is another question.

Many religions have a myth that it existed forever. Obviously, those religions are of human origin. The all-knowing extra-terrestrial intelligence (the one true God) did not make that mistake.

The bible, and science both say that it had a beginning.

---

[12] Complete outline is found in Appendix C:

28

It has not been that many years ago that any self-respecting scientist could say that the bible is false because it makes the claim that the universe had a beginning. At that time the accepted theory was that the universe was "steady state" that is, had existed forever. There was no beginning. It has always been. Ironically that religiously held "scientific" view agreed with other, obviously man made, religions.

Any man made religion of course, would be expected to reflect the opinion that the universe had always been. But God did not make that mistake.   Against all opinion of humanity, against political correctness, against science by consensus, the bible openly declared that the universe had a beginning.

And then modern science discovered the bible was correct on that point.

But the bible recorded that point of detail thousands of years before its discovery by modern science.

Does that mean the bible has a supernatural origin?

That is a good start, but by no means all of it. That is just scratching the surface of what modern science has independently discovered, but the bible said first—by thousands of years.

In the contest of who said it first, the bible is going to win overwhelmingly.

Currently, there are various theories as to what actually happened in the origin of the universe.   Most theories involve the sudden expansion from a point, possibly a "black hole," or at least "darkness" of some kind, and then the sudden release of light along with all the other aspects of the universe.

All theories seem to agree on several points of detail.

The universe really did have a beginning.

Before that beginning, the universe did not exist as we know it today.

When the universe began, energy in the form of light was released.

That light originated, or was released, from some form of darkness, possibly a black hole.

The beginning of the universe was an, explosive event.

Among the current theories, one of the oldest is recorded in the ancient scriptures of the bible. Each and every one of the above points of detail is included in the biblical accounts, including the points that it was an explosive event, and that light came from darkness, not from nothingness.

Modern science agrees with the bible on every one of the above points.

The Science

The theory as to the events involved in the origin of the universe that enjoys the most widespread awareness among the general public is the one the scientists call the Big Bang Theory.

According to the Big Bang Theory:

> The universe started at a specific time and place with a great explosive expansion. Possibly it was an explosion of a pre-existing black hole.
>
> The universe is the fourfold interrelationship among matter, energy, space, and time. Without any one, the others do not exist.
>
> There was an event that simultaneously began all the relationships.
>
> Related to that event, light was one of the first things to be released into outer space.
>
> The beginning was a real event.
>
> The beginning was not only an event; it was a fundamental change from the lack of existence of what we know today as the universe, or cosmos.
>
> Sometime in the far past there was a catastrophic event which eventually resulted in the existence of what we know today as the universe, or cosmos.

Another, more recently published theory[13] is the "Big Bounce" theory. It assumes this occurrence is simply a repeat of previous

---

[13] *Big Bang or Big Bounce?: New Theory on the Universe's Birth.* Our universe may have started not with a big bang but with a big bounce—an implosion that triggered an explosion, all driven by exotic quantum-gravitational effects. Scientific American Magazine, October 1, 2008 Page 40. Retrieved from the internet October 1014 at http://www.scientificamerican.com/article/big-bang-or-big-bounce/

occurrences where the universe, simply explode from a black hole, become a universe, and eventually re-collapse back into a black hole state where it again explodes into another full blown universe.

Other theories include mechanisms to set up the conditions for the current universe to come into existence. What it was that existed before the big event that started the current universe is still a topic of hot debate. The one thing that is no longer debated is the fact of a beginning.

In 1963, Arno Penzias and Robert Wilson discovered the cosmic microwave background radiation of the Big Bang Theory. This was accepted by modern science as convincing evidence that the universe was born at a definite moment. For this discovery, they were awarded the Nobel Prize in Physics in 1978.

So, the concept of the universe having an actual beginning is a recent discovery of modern science. But the bible published it first by thousands of years.

The Bible
This second eon began with the coming into existence of the universe, the creation of the cosmos,
 As the bible puts it:
   In the beginning God Created the Heaven…
     Moses, c. 1445 BC, *Introduction to Creation II,*
       A Biblical Creation Account: Genesis 1:1,2
A scientist might say, it happened with the *Big Bang.*
The bible says it happened with a *Powerful Blast.* Observe the following account with a reverse chronology going back to the very beginning where it is described as a "blast."
 4 Where wast thou when I laid the foundations of the earth?
  declare, if thou hast understanding.
 5a Who hath laid the measures thereof, if thou knowest? or
 5b Who hath stretched the line upon it?
 6a Whereupon are the foundations of the foundations fastened? or
 6b Who cast out the corner stone thereof;
 *7a When the first generation stars were **blowing out** together,*
 7b And all the members of God

    [7c]*gave a blast?*
                        God, c. <1500 BC, *Introduction to Creation I,*
                            A Biblical Creation Account: Job 38:1-11

   A Godly Blast, a Big Bang, these are just slightly different words for the same event.   But the bible published it first, by several thousand years.

A Question for Both Science and the Bible: What Caused It?

   The answer to that question is outside the realm of the scientific method.

   The most universally recorded detail of all the creation accounts of the bible, is the fact that it was God that did the creating, and that which God planned and designed was actually brought into existence. That claim is recorded in every biblical creation account.

   But modern science has expressed the opinion that it had to arise by spontaneous generation.

2.2. The Expansion of the Universe

The Expansion of the Universe is the second phase in Eon 2, the Eon of Early Development.

2.2.   The Expansion of the Universe
2.2.1   Then there was the expansion of the universe.
2.2.2..   Then water became abundant in outer space
2.2.3..   Then stars came into existence
2.2.4..   Then the Sun came into existence

The Science

The Expansion of the Universe:

It has not been very many years since Albert Einstein (1879-1955) published the theory of relativity, (1915/16) which establishes the relationship between matter, energy, space, and time. At that time the prevailing notion was that the universe was static, not expanding nor contracting. Einstein was dismayed to find that his theory predicted either an expanding or contracting universe.

In what he later called "the greatest blunder of my life," Einstein added a "cosmological constant" to the equations that would make his math agree with a static universe.

It was in 1929 when Edwin Hubble (of the Hubble Telescope) showed that distant galaxies were receding from the earth and the further away they were, the faster they were receding. It was then; in 1929 that Einstein discovered his blunder. That discovery changed cosmology.

It was less than a hundred years ago that modern science discovered the universe was expanding. The ancient scripture recorded that detail of science thousands of years earlier.

Water, Then Stars, Then the Sun

Even as this book is being written, modern science is learning about water in outer space and its relationship to the formation of stars, the sun, and planets. In 2001, an article[14] by the staff at SPACE.com posted on the Internet included the following statements:

---

[14] *http://www.space.com/scienceastronomy/astronomy/milkyway_water_010412.html*

"New observations of selected regions of our Milky Way Galaxy show that water is more abundant than expected. The new measurements show that water is the third most common molecule in the regions studied, giving researchers useful information about the abundance of elements available when new planetary systems are formed."

"Using European Space Agency's Infrared Space Observatory, Spanish and Italian astronomers have for the first time measured the amount of water in cold regions of our galaxy. This is especially interesting, the researchers say, because these regions are the birthplace of stars like the Sun and, in some cases, the planets that form with them. The mean temperature of the water in these cold regions is minus 441 degrees Fahrenheit (minus 263 degrees Celsius)"

No wonder the translators of the bible had difficulty and ended up with theobabble when confronted with these concepts. These concepts were not discovered, not even imagined, until thousands of years after the ancient scriptures of the bible were written—even thousands of years after the original Hebrew language became a dead language and the meaning of many words forgotten—and hundreds of years after the first English language translations set the tradition of what the original language presumably meant.

But isn't this discovery what we would expect when we get down to the simple truth of what was recorded by Job and David thousands of years ago.

From the beginning, the universe expanded from a central location.

The universe continues to expand from that point of origin.

As time advanced from the beginning, there was an orderly sequence of events in the formation of the galaxies.

That sequence of events contained several stages of development

In one of the earlier stages water became abundant, and remains abundant in outer space to this day.

According to Modern Science, Hydrogen, H_2, is the most abundant element in the universe.[15] And it was so from the first elements to

---

Water in Space More Abundant than Expected, by SPACE.com Staff, posted 04:00 pm ET, 12 April 2001

[15]*David Palmer, for Ask an Astrophysicist, in answer to a question submitted November 13,*

develop in the universe. Water, H_2O, is among the most abundant molecules in the universe.[16] And water began to accumulate before the current generation stars came into existence. In the free space of the universe is a vast amount of water

There are at least three generations of stars.

In one of the later stages, the star we call our sun, a third generation star, came into existence.

Yes, modern science affirms the chronology is correct.

The Bible says:

Thousands of years ago, when speaking of the greatness of God and listing God's creation feats, Job published the following chronology:

> [8] *Which alone spreadeth out the heavens,*
>> First there was the expansion of the early universe.
>> *and treadeth upon the heights of the sea.*
>> Then water was abundant in the early universe.
> [9] *Which maketh Arcturus, Orion, and Pleiades,*
>> Then came the stars and constellations
>> *and the planets of the southern sky.*
>> And finally the planets of our solar system.
>> Job, c. <1500 BC, *The God Alone Creation Account,*
>> *A Biblical Creation Account:* Job 9:8,9

Did Job get the chronology correct? Modern science affirms it.

A few hundred years later, King David recorded a chronology of creation events in the midst of a praise of the greatness of God:

The Cosmos.

[2a] *Who coverest [thyself]*
God was the cause of light (the universe) coming into existence.
[2b] *with light as [with] a garment:*

1997, Accessed on the internet March 18, 2007, *http://imagine.gsfc.nasa.gov/docs/ask_astro/answers/971113i.html* "When the Universe was formed in the Big Bang, the resulting elemental matter was about three quarters hydrogen, one quarter helium, and a few parts-per-billion of lithium (by weight)... Most of it is still around, and so the elemental matter of the Universe is still about three quarters hydrogen,..."

[16] *http://en.wikipedia.org/wiki/Water_%28molecule%29#Water_in_the_Universe from Wikipedia, the free encyclopedia:* Water (molecule): Water in the Universe, *Accessed March 18, 2007,*

Light was one of the first things to appear at the beginning..

Light had a beginning. It has not shone forever.

2c who stretchest out

Then there was the expansion of the universe.

2d the heavens (cosmos) *like a curtain:*

The heavens (including water as implied by the mention in next verse, and that which is seen: (the sun and stars) came into existence after the light and expansion.

Water was abundant in outer space.

The stars came into existence.

The Sun came into existence.

The Water in The Cosmos.

3a Who constructs his chambers of heaven (planets)

Then the solar system developed.

The earth and other planets developed.

The planets are chambers for storage of water to moderate the water cycle in outer space.

3b in the waters [of outer space]:

Water had accumulated in solar system.

The planets were formed in outer space where water vapor had accumulated.

David, c. 1015 BC, *Chronological Order of Creation, A Biblical Creation Account:* Psalms 104

Did David get the chronology correct? Modern science affirms it.

In the next example, David is praising God and gives another chronology. This time it is in reverse chronological order:

(Order of phrases is presented in reverse to make chronology flow in positive direction.)

4b[Praise him,] ye waters that [be] above the atmosphere

The fourth item mentioned in this chronology is the water of outer space.

The elements of water developed first after the beginning and before the stars.

4aPraise him, ye heavens of heavens,

The whole cosmos declares the glory of God.

3bPraise him, all ye stars of light.

The fifth item mentioned in the chronology is the stars.

3aPraise ye him, sun and moon:

The sixth item mentioned is the sun.

The seventh item mentioned is the moon.

No mention has been made of the continents up
to this point, as they had not yet formed.

David, c. 1015 BC,

Before and After Account,

A Biblical Creation Account: Psalms 148

Again, did David get the chronology correct? Modern science
says yes. And this is only a partial list of ancient scriptures giving
correct chronology.

Chronology of Creation Events:

8   Which alone spreadeth out the heavens,

Expansion of the universe.

and treadeth upon the heights of the sea.

"The heights of the sea" is a reference to the water of outer space.

9   Which maketh Arcturus, Orion, and Pleiades,

Development of the Stars and Constellations.

and the planets of the southern sky.

Development of the Solar System.

Job, c. <1500 BC, *The God Alone Creation Account,*

A Biblical Creation Account: Job 9:8,9

Birthplace of the Solar System.

3a Who constructs his chambers of heaven (planets)

Water existed in outer space before the planets formed.

3b in the waters [of outer space]:

David, c. 1015 BC, *Chronological Order of Creation,*

A Biblical Creation Account: Psalms 104

Not only did the ancient scriptures get the chronology correct, it got
the detail that the third generation stars, our sun, solar system with
planets, were all birthed in water rich areas of outer space.

And modern science was not fully convinced until after the writing
of this very book you are reading, was underway.

Note inserted September 26, 2014:

Today, as I do the final editing of this section, I received an email from Clyde[17] alerting me of new research announced yesterday, Thursday, September 25, 2014 in a Reuters news release[18], *Study finds solar system's water older than the sun..* Lead researcher, Lauren Cleeves, a doctoral student at the University of Michigan claims to have discovered that Water found in Earth's oceans, in meteorites and frozen in lunar craters predates the birth of the solar system,

The article claims this is significant in the search for life on other planets. More significant is the fact that this confirms the chronology information published in the bible thousands of years ago is in accord with reality that was not discovered by modern science until now.

But isn't this what you would expect if this book contains any truth?

---

[17] *Clyde Spencer, major critic of my work, long time friend, recognized as the most intellectually honest person I know.*

[18]http://news.msn.com/science-technology/study-finds-solar-system%e2%80%99s-water-older-than-the-sun *and*
http://www.ns.umich.edu/new/releases/22401-the-water-in-your-bottle-might-be-older-than-the-sun

2.3. The Development of the Solar System

The Development of the Solar System is the third of six phases in Eon 2, the Eon of Early Development.

This phase has two major developments. In the first, the planets them selves are formed, including the planet earth. In the second, a water cycle develops in the outer space of the solar system whereby water is carried out towards the outer reaches by solar winds, and returned back toward the center by gravity. A visible glow is a characteristic of the solar winds.

2.3. The Solar System Develops- *The Chambers of the South*
2.3.1. Then the solar system developed
2.3.1..1. The earth and other planets developed.
2.3.1..2. The planets are chambers for storage of water to moderate the
              hydrologic cycle of outer space
2.3.1..3. Water had accumulated in solar system
2.3.1..4. Before life on earth
2.3.2. Then hydrologic cycle (water cycle) developed
              in outer space of the solar system.
2.3.2..1. That outer space hydrologic system is permanent
2.3.2..2. Intelligently designed for a purpose,
              To fill and maintain future oceans of planet earth.
2.3.2..3. Water carried in by the forces of gravity (small comets?)
2.3.2..4. Solar winds existed in solar system
2.3.2..4.1. Water carried out by solar winds.
2.3.2..4.2. Visible glow is a characteristic of solar winds

Where does the bible say anything about the development of the solar system? Where does it mention the planets? For a long time, that was a puzzle.

Another puzzle concerned the meaning of the reference in Job[19] to "the chambers of the south." What was it talking about? I remember a long time ago, back in the 1960s, my bible college professor saying something about the theologians in their current state of understanding about that reference. He said no one knows what the phrase, "the chambers of the south" is referring to. Then, one day I

---

[19] *Job 9:9* Which maketh Arcturus, Orion, and Pleiades, and the chambers of the south

    39

was working on the puzzles of the conversations between Job and God about the topics of science, when I happened to be looking at the simulation of the night sky that came with my telescope. All of a sudden it made sense. The listing of heavenly bodies of the night sky there in Job around that phrase was missing only one category of objects—the planets. From the point of view of the northern hemisphere, where Job must have been observing the night sky, the planets all are in the southern sky. There it was. That is where the bible refers to the planets of our solar system. The word for planets has been traditionally translated into the word, "chambers" So I looked at other references to "chambers" and found in another biblical creation account[20] a reference to the construction of the "chambers" right where one would expect the sequence of events to be talking about the development of planets in the solar system. In another reference[21] where it is talking about sustaining the water supply of our planet to maintain the ecology, it talks about a source of sustaining water being from storage areas on other planets called chambers.

The Science

According to the Modern Science, the solar system developed in a water rich region of outer space.

The Solar Wind

Hydrogen, H_2, is the most abundant element in the universe.

The solar wind is fed by hydrogen from the sun.

Humans have observed the effects of the solar wind since the beginning of mankind in the glow of the aurora borealis and the fiery glow of the tails of comets.

But their scientific explanation was not discovered by modern science until 1958. In the 1600's Kepler guessed that the pressure of sunlight drove those tails. In 1953, Cuneo Hoffmeister in Germany, and later, Ludwig Biermann, proposed that the sun emitted a stream of particles. But it was not until 1958 that Eugene Parker of the University of Chicago, could explain the solar wind.

---

[20] *Psalms 104:3* "Who layeth the beams of his chambers in the waters:…" *refers to the beginning of construction (laying the beams) of the planets (chambers) in a water rich region of outer space.*
[21] *Psalms 104:13.* "He watereth the hills from his chambers…"

The Water Above:

Water, H_2O, is among the most abundant molecules in the universe.

In the free space of the universe is a vast amount of water

The planets of our solar system are and have been in the past, vast storage reservoirs of water.

The oceans of the earth, the mysterious missing oceans of the past of Mars—that are still baffling the scientists of NASA—and the Ice fields of other planets, all are, and have been, vast storage chambers of outer space water.

In 1986 Louis A. Frank of the University of Iowa, published his theory of small comets. His theory contends that there are multitudes of small comets in outer space. These small comets are best described as giant snowballs with a size on the order of the size of a house. That publication claims they discovered hundreds; even thousands of these small comets are being captured each day by the atmosphere of the earth, being disintegrated by the atmosphere, eventually falling to earth as rain. The amount of rainfall attributable to these incoming small comets is sufficient to have filled the ocean several times over.

Whether or not small comets are the proper explanation, the influx of extraterrestrial water provides the answer to the lingering question scientists have had for decades. Why has the water on the earth not escaped to outer space? Why has the water of the earth not been blown off by the solar wind and the oceans of the earth been lost to outer space? We see water being blown off comets by the solar wind. That is how we explain the tail of comets. The answer has become obvious. The water has been blowing off the earth, just as it has been blowing off all other heavenly bodies. The incoming extraterrestrial water being captured by our atmosphere is simply replenishing it.

From this point, it is only a small leap of hypothecation for scientists to come up with the theory of the hydrologic cycle in outer space. According to this hypothesis, water is falling in toward the sun (possibly in the form of small comets.) As the water approaches the sun it is heated and disassociated into ions that are carried back out by the solar wind. Water is going out via the solar wind, recombining, and returning back in, possibly by the small comets.

Just as the theory of the geocentric universe has gone by the wayside, another hypothecation by modern science will destroy another religiously held theory. If the theory of Small Comets is true, there is no Oort Cloud. The origin of comets, all of which are very young compared to the universe, needs no further explanation than the occasional deflection of one of these small comets by a close miss with a planet. By being deflected just the right amount, a small comet will be set into elliptical orbit around the sun where it will go back out to the outer reaches of the solar system and return some day with added mass of water and space debris. It will become a full-fledged comet and eventually die the death typical of all comets.

To Copernicus goes the credit for the discovery that the planets revolve around the sun. However, it is difficult to give credit to the scientist who discovered that the planets are storage chambers for water in the solar system hydrologic cycle. That discovery is too recent for the scientific community to have realized that it has been discovered.

In 2006, American scientists were amazed[22] to discover a volcano on one of the moons of one of the other planets spewing liquid water. The Old Scientist would propose the credit for the discovery that the planets are storage chambers for water should rightfully go to the author of the bible.

The Bible

According to the Ancient Scripture:

In the *Chronological Order of Creation* account, assumed[23] to be put into the biblical record by King David about 1015 BC, we find the following description of the development of the solar system, including the planets and the hydrologic cycle in the outer space of

---

[22] *Baker, Joanne, 2006, Tiger, Tiger, Burning Bright, Science 10 March 2006: Vol. 311. no. 5766, p. 1388 DOI: 10.1126/science.311.5766.1388*
    http://www.sciencemag.org/cgi/content/short/311/5766/1388 or:
Connor, Steve, Science Editor, 2006, Water plumes spewing from 'ice volcano' seen on a moon of Saturn, The Independent, Online Edition, Science & Technology, March 10, 2006.   http://news.independent.co.uk/world/science_technology/article350375.ece retrieved April 25, 2007.
[23] See, *The Mystery of Psalm 104*, page 27, *The Mystery of The Charge of Plagiarism*, for an analysis of Psalm 104 and the Hymn to Aten, found at *www.EyeWitnessToTheOrigins.com*

the solar system. It appears to be in the appropriate chronological position. It appears after the expansion of the universe and before the molten surface of the planet earth that would later develop into the foundation of the continents.

For over two thousand years this passage of scripture was translated into some fanciful theobabble about angels, spirits, and God walking on the wings of the wind. Little did the theologians realize they had no idea of what this ancient account was all about. All the words used, the immediate context, and the fact that this psalm is a listing of the overall chronology of existence, all cry out that this is in fact a description of what modern science is in the process of discovering even as this book is being written. The planets of this solar system were birthed in a water rich region of outer space, and there is a hydrologic cycle within the outer space of the solar system.

Development of the solar system in the water of outer space.

3a Who constructs his chambers of heaven (planets)

> Then the solar system developed.
> The earth and other planets developed.
> The planets are useful as chambers for
> storage of water to moderate the cycle.

3b in the waters [of outer space]:

> Water had accumulated in outer space.

Setting up of the hydrologic cycle in the outer space of the solar system.

3c who set up the dark clouds (extraterrestrial water)

> Hydrologic cycle (water cycle) developed
> in outer space of the solar system

3d to be his [water] conveyors:

> Water carried in via extraterrestrial Hydrologic cycle

3e who carries [water]

> Water carried out by solar winds.

3f upon the far reaches of the winds [winds of heaven] (Solar Winds)

> Solar winds existed in solar system

4a Who maketh his messengers winds [of heaven] (Solar Winds);

> Solar winds are used to transport water

---

   ⁴ᵇ his servants a flaming fire:
> Visible glow is a characteristic of solar winds
> (as in the tail of a comet as water is blown off,
> away from the sun.)
> David, c. 1015 BC,
> *Chronological Order of Creation,*
> *A Biblical Creation Account:* Psalms 104

In *The God Alone Creation Account,* Job included a chronological order list of events placing the origin of the planets in the proper position. The Solar system came after the development of the stars and constellations. These came after water was abundant in outer space, after the heavens were being spread out.

Job's Chronology of Creation Events:
⁸ Which alone spreadeth out the heavens,
> First there was the expansion of the early universe.
and treadeth upon the heights of the sea.
> Then water was abundant in the early universe.
⁹ Which maketh Arcturus, Orion, and Pleiades,
> Then came the stars and constellations
and the planets of the southern sky.
> And finally the planets of our solar system.
> Job, c. <1500 BC,
> *The God Alone Creation Account,*
> *A Biblical Creation Account:* Job 9:8,9

In the *Eyewitness Account by Wisdom,* recorded in the book of Proverbs about 970 – 930 BC, Wisdom specifically states that God planned the oceans, and the water of outer space was part of that plan.

²⁷ᵃ*When he prepared the heavens, I [was] there:*
> Here Wisdom begins a second pass through the chronology. The first pass was more a statement of what was. This second pass is more of a statement of what God did about it. The preparation of the heavens is another phrase describing the fact that, as modern science has speculated, the heavens were prepared, planned,

setup conditions, before the universe came into existence. At that stage of preparation, Wisdom was a participant. Wisdom is the designing engineer aspect of God.

27b*When he set a compass upon the face of the ocean:*

The oceans were planned, and God did the planning. This was the intelligence (Understanding, Logos, Word) aspect of God as described in other accounts.[24] [25]

28a*When he established the clouds above:*

This has been interpreted to be a reference to the establishment of the hydrologic cycle on the planet earth. However, that occurred in a later era than this immediate context. Instead, this is a reference to clouds of water in outer space from which the oceans were filled, as recently discovered by modern science. Those clouds formed in the Pre-Hadean Era, the context of this account. Most likely this is a reference to the establishment of the atmosphere that captured water from outer space to fill the oceans.

Wisdom, c. 970 – 930 BC,
Eyewitness Account by Wisdom,
A Biblical Creation Account: Proverbs 8:1,12,22-31

---

[24] *John: c. 80-95 AD.* The Logos Creation Account, *"In the beginning was the Logos, and the Logos was with God, and the Logos was God. All things were made by him..."*

[25] *Jeremiah: c. 585-580 BC.* The Triune Creator, *"He hath made the earth by his power, he hath established the world by his wisdom, and hath stretched out the heaven by his understanding. "*

2.4. The Development of The Planet Earth

Note: In reading this, remember the word "earth" in the bible does not refer to the planet earth. "The earth" is the continental part of the planet and "the deep" is the oceanic part.

The Development of the Planet Earth is the fourth of six phases in Eon 2, the Eon of Early Development. In the first part of this phase, before the surface cooled sufficiently for water to accumulate, there are several details listed as operating together with no chronological significance. Remember, in the numbering system of the outline, the numbers after the double decimal have no chronological significance relative to each other, only that they occur in the same era as indicated by the numbers to the left of the double decimal.

2.4. The Development of the Planet Earth

2.4.1..   Then planet earth developed with molten magma surface (mantle.)

2.4.1..1   Mantle Planned to support the earth (continents.)

2.4.1..2   Mantle Engineered to support the earth (continents)

2.4.1..3   Molten mantle develops, to support the earth (continents) (the "molten support" erroneously translated "pillars.")

2.4.1..4   Solidified Mantle is to become the ocean floor

2.4.1..5   Held in place by gravity/density (floatation)

2.4.1..6   Without continents

2.4.1..7   And without soil

2.4.1..8   Without oceans.

2.4.1..9   Without surface water

2.4.1..10 No humans existed

2.4.1..11. With Juvenile water coming from inside the earth

2.4.2..   After a while the planet earth cooled and developed surface water

The early development of the planet earth was quite a process. For it to end up with vast oceans surrounding major land masses called continents took several processes in the proper sequence. For the fortuitous sequence of events to come off without a hitch seems almost like it had to have been engineered by some supreme intelligence with the power to make it happen. But that is the domain of the bible, not of science.

Why is a "Foundation" for the continents Significant?

The surface of the planet earth is mostly covered by water. Over two thirds (70.8%) of the surface of the planet earth is covered by oceans. Less than one third (29.2%) is land. The ocean is an average of almost two and a half miles deep.[26] The average height of the land of the continents that sticks up out of the water is only about a half mile. That means the continents are piles of rock about three miles high compared to the ocean floor.

But it wasn't always that way. In an earlier stage of development, the planet earth did not have an ocean at all. Then, after that, it was all covered by ocean with no continents at all.

Imagine, if the continents were spread out over the whole surface of the planet, the ocean would still be about two miles deep over all of it. There has to be some mechanism keeping the continental rocks piled up high enough to stick out above sea level.   Otherwise, erosion would wear the continents down to sea level.

How did those rocks get piled up in the first place? And what is beneath the continents that holds them up? In other words, what is the "foundations of the continents?" Or in the language of the bible, the "foundations of the earth"—remember, the bible uses the term "earth" to mean the land mass part of the surface of the planet, not the planet itself.

To make it even more interesting, somewhere in the early development of the planet earth, scientists have the theory that the moon came from the surface material that had been an outer layer of the planet before something catastrophic happened. At a particular stage in the development of the planet earth, after it was a spherical mass of molten rock, after the layers of rock material had separated by density, causing the lighter rocks to be closer to the surface, something catastrophic happened that changed the whole course of development of this planet. The whole earth, not just a small fraction as it is today, was covered with material similar to what the continents are made of. At that point in time, there was not much of a chance of the modern ocean floor of today being exposed to the water at the

---

[26] *PhysicalGeography.net | FUNDAMENTALS eBOOK CHAPTER 8: Introduction to the Hydrosphere, http://www.physicalgeography.net/fundamentals/8o.html*

bottom of the ocean. You see, the earth is made up of concentric layers of material. Starting at the core, the material is very dense (heavy) with each concentric layer, as you work your way to the surface, being less dense than the one under it. This results in a stable situation where no lighter surface layer can sink into the denser layer below.

At that time, before that catastrophic event that tore off that great quantity of continental material to make the moon, the outer layer of continental material was so thick that there was no mechanism that could pile up part of it to be much higher than the rest of it so it could be above sea level. You see, the ocean is on the average, about two and a half miles deep. Something had to happen to make some of the continental material stand over three miles higher than the ocean floor so the ocean water would have a basin to run off in to. Otherwise, the whole surface of the planet would be covered with ocean water and we would have no continents.

But then some catastrophic event did occur at just the right time in the development of the planet. After the lighter continental rock material had come to be the outer layer of rock, another heavenly body about the size of the planet earth collided with the earth and stripped much of the continental material off into outer space. That, scientists say, is where the material came from to make the moon.

But the more significant result of that catastrophic event was the exposing of the more dense layer that was below the surface. With that more dense layer, later to become the foundations upon which the continents stand, exposed to the cold water of the ocean, the ocean floor became a solid layer of the same material that lies below it. That caused a density inversion where more dense rock is over and on top of less dense rock such that if it should break, it would sink into the depths of the molten rock below. And break, it does. That layer, about five miles thick, overlying the "Moho: as scientists call it, is broken off from the similar layer under the continents. This has occurred at the edge of the continents where it sinks under the continent. This sinking of the ocean floor at the continental the edge is know to scientists by the term, "subduction." Remember that term subduction as you will see it again in this writing. As it sinks, it slides sideways towards and under the continents, re-depositing the weathered continental material that has spread out over the ocean floor. That mechanism re-deposits that re-cycled continental material back under the roots of the continents, causing the continents to be buoyed up by the lighter continental material under the roots of the mountains. This causes the continents to stand up in piles over three miles high, to be above sea level, and to remain above sea level, in spite of the erosion that tends to wear them down.

The Bible

Foundation of the earth

I guess, if you believe all this stuff the scientists talk about, it looks like it was planned. You might think all that complex manipulation of the surface materials of the planet is the result of engineering the molten mantle to be the foundations of the continents.

In the *Chronological Order of Creation* account recorded in the bible around 1015 BC, probably by King David, the liquid mantle surface is mentioned as being laid on the developing planet earth. That account mentions it just after the description of the development of the hydrologic cycle in the outer space of the solar system, and before mentioning the filling of the ocean. The same account also mentions that it would later develop into the foundations that hold the continents up by describing that process as the laying of the foundations of the continents to hold them in place.

Historically, this is one of the scriptures that were misinterpreted to convict Galileo of heresy. The theologians of that time interpreted holding the "earth" in place to mean the planet earth could not possibly be rotating around the sun. It should have been interpreted to say that since the continents were made of less dense rock material than the foundation material, the continents would not sink into the interior of the earth. But then, that scientific detail was not discovered until long after Galileo was convicted.

> [5][Who] laid the foundations of the continents,
> (laid the liquid mantle)
> [that] it (continents) should not be removed for ever.
> David, c. 1015 BC,
> *Chronological Order of Creation,*

In the *Introduction to Creation I* account, recorded in the book of Job long before 1500 BC, God in person asked questions indicating God understood that during the construction of the planet earth, the mantle was laid out before the continents existed. The questions also indicate God understood it was planned to support the continents.

> ^{4a} *Where wast thou when I*
>> Assuming here that God is asking about the origin of the universe, it is obvious that he is saying there was a time when the mantle layer of the planet earth did not exist, and then it did. It developed some time in the past.
>
> ^{4b} *laid the foundations of the earth?*
>> That which holds up the continents is the mantle, laid out during the formation of the planet earth. Of course, it is obvious that at this time there were no continents. They had not yet formed.
>
> *declare, if thou hast understanding.*
> ^{5a} *Who hath laid the measures thereof,*
>>> Who was the scientist who planned it?
>
> *if thou knowest? or*
> ^{5b} *Who hath stretched the line upon it?*
>> Who was the engineer that designed it?
>>>> God, c. <1500 BC,
>>>> *Introduction to Creation I,*
>>> *A Biblical Creation Account:* Job 38:1-11

The Eyewitness, in the *Eyewitness Account by Wisdom,* tells of a time then there were no continents and no oceans. Notice this is a reverse chronology, going back in time from the emergence of the continents from the oceans, through the time before there were no oceans, back to the time even before there was liquid water coming from the interior of the earth. Early in the development of the planet earth, there were no continents, there were no oceans, there were even no water spewing out of volcanoes. It must have been too hot for water to be liquid. That is when the rock layer called the mantle, which the bible calls the foundations of the earth, meaning the foundations of the continents, was laid.

23a *I was set up from everlasting,*

23b *from the beginning,*

23c *or ever the earth was.*

> The meaning of this strangely worded phrase is, "before there was any earth (continents.)" This means there was a time when there were no continents at the particular stage of development that scientists call the Early Hadean Era. Long before the emergence of the continents (earth.)

24a *When [there were] no depths, I was brought forth;*

> It was even before there were any oceans.

24b *when [there were] no fountains abounding with water.*

> It was even before there were any volcanoes
> spewing out water before the oceans were filled.
>
> Wisdom, c. 970 – 930 BC,
> *Eyewitness Account by Wisdom,*
> *A Biblical Creation Account:* Proverbs 8:1,12,22-31

The Science on Early Surface Water

Eventually the surface cooled sufficiently for liquid water to begin to form on the surface: Before the filling of the oceans, juvenile water spewed out from the interior of the earth sufficiently to wet the whole surface. There was a time that it had not rained. It was the time before the atmosphere developed. There was no atmosphere to carry water from pooling areas to dry areas and then drop it from the air. Eventually, water and gasses, (vapor) spewing from the interior of the planet through volcanoes developed into atmosphere and a hydrologic cycle developed, but that happened later. We are getting ahead of ourselves in the story.

The Bible on Early Surface Water

5c   [At a time when] the LORD God had not yet made
the rain on the earth

5d   and there existed no human to work the ground,

6a   vapor spewed up from the earth,

6b   and caused water

6c   on all the panorama of the ground (whole surface.)

Moses, c. 1445 BC, *One-Day Creation Account,*
A Biblical Creation Account: Genesis 2:4-7

More on the Forming of the Foundation of the earth

Having been brought up in Sunday school and having been told the story of creation many times from the point of view of the traditional interpretation of the "official version," it is hard to believe that the bible actually teaches what modern science has discovered.

According to most any science reference readily available, scientists assume the earth at one time was hot and had a molten surface. However, no one seems to want to claim credit for that discovery. The assumption is that it cooled enough for water to accumulate. But no one wants to claim credit for that discovery. Apparently, modern science assumes it is so obvious that the earth, in ancient times, went through phases of being too hot for water to be on the surface, and that volcanic action was the source of the first water on the surface when it finally cooled sufficiently. No one is given credit for discovering the obvious.

However, not many years ago, no one knew of this. It was assumed that the earth was created cool as it is today, and in appearance as we see it today.

At least, the bible has a position on some of the details, and they were published thousands of years ago, and they seem to be in accord with the latest views of modern science, and they seem to be at odds with the traditional interpretation of the bible on these science related subjects. Thus, the bible is much better than any religion, including its own. Many religions, including the one based on the bible, have made up ridiculous explanations that are obviously not true to facts of science, nor true to the biblical accounts.

To a scientist, the idea of laying the foundation of the continents may seem like a ridiculous thing to mention because it is so obvious that something holds up the continents. However, a deeper study into the words used in the ancient Hebrew language brings the realization that God knew more about it than appears on the surface. God knew the foundations of the earth were the molten support for the continents. That "foundation" is known to science as the mantle. God also knew it sometimes leaks to the surface as lava flows.

> *"...for the pillars of the earth [are] the LORD'S, and he hath set the world upon them."*
>
> Samuel, c.925 BC, *Concept of Universe,*
> *A Biblical Creation Account:* I Samuel 2:8

In the original Hebrew language, the word translated "pillars" means "molten support."

How could something that meant "molten support" be translated "pillars?" If one did not realize that King David understood that melted rock was the molten support for the continents, you might miss the significance of what he said when he saw a river of hot lava leaking from beneath the continent.

King David seemed to be quite affected by an experience he had on the occasion that he was spared from his enemy by a volcano erupting and spewing smoke, fire, hailstones and falling hot lava. He even saw the "foundation of the earth" (melted hot lava) flowing out onto the surface. His description is recorded in multiple places in the bible, both in the book of 2 Samuel, and as the eighteenth Psalm. He summed up the occasion towards the end of the account with this description:

10 *...And he rode upon an erupting volcano, and did fly: yea, he did fly upon the wings of the wind.*
11 *He made darkness his secret place; his pavilion round about him [were] dark waters, thick clouds of the skies.*
12 *At the brightness before him, his thick clouds passed, hail and coals of fire.*
13 *The LORD also thundered in the heavens, and the Highest gave his voice; hail and coals of fire.*

14 *Yea, he sent out his arrows, and scattered them; and he shot out lightnings, and discomfited them.*

15 *Then the channels of waters were seen, and **the foundations of the world were discovered** at thy rebuke, O LORD, at the blast of the breath of thy nostrils.*

16 *He sent from above, he took me, he drew me out of many waters.*

17 *He delivered me from my strong enemy, and from them which hated me: for they were too strong for me.*

18 *They prevented me in the day of my calamity: but the LORD was my stay...*

David, c.1015 BC, *Deliverance From My Enemy*,
A Biblical Creation Account: Psalms 18

Notice the phrase in bold. After the falling hot lava fell from the sky, David observed molten lava that he described by a term familiar to him—the foundations of the earth.

Also notice that the translation of David's description of being rescued by an erupting volcano is quite cumbersome because the original translators did not even recognize it to be what it was, an erupting volcano.

2.5 & 2.6 Development of the Atmosphere &Oceans

The Development of the Atmosphere and the Oceans is the fifth and sixth of six phases in Eon 2, the Eon of Early Development.

The bible makes a big deal of these two processes and their order of occurrence. Biblical accounts describe in detail the sequence of events in the development of the atmosphere and the development of the oceans. Apparently, the atmosphere had a part in filling the oceans so it had to develop first. And that is the way it is presented in the biblical accounts, all in accord with reality, not religion.

2.5.   The Development of the Atmosphere
2.5.1..   Then an atmosphere developed.
2.5.1..1.   Atmosphere captures water from outer space.
2.5.1..2.   Watered the surface of planet earth
2.6.   The Development of the Oceans - *The Deep*
2.6.1..   Then the atmosphere captured (separated, set apart) water from
              outer space.
2.6.1..1.   Adding to original juvenile water
2.6.1..2.   Filling the oceans
2.6.1..3.   Separating (extracting) water below the atmosphere (oceans)
2.6.1..4.   (direction of separation (extraction) was from above to below)
2.6.1..5.   from water above the atmosphere (small comets.)
2.6.2..   Then the earth was covered by the oceans,
2.6.2..1.   Solidified mantle became ocean floor
2.6.2..2.   Intelligence existed prior to this
2.6.2..3.   Engineering existed prior to this
2.6.2..4.   No continents existed at this time.

The Earth with Oceans—but without Continents.

Scientists' attitude on earth covered by ocean

It appears to be commonly accepted by earth scientists of today that there was a period of time in the far past of the history of the development of the planet earth, during the Hadean Eon, in which the planet earth was covered with ocean and without continents.

Although The Old Scientist has not yet been able to find to whom to give credit for that particular discovery, it appears to be of recent history. The basis given for some to come to that conclusion is the

55

virtual absence of anything except pillow basalt in the earliest rocks of the geologic column. Pillow basalt is unique in its formation in that it is formed under water when hot melted lava is rapidly cooled by water, and forms pillow shaped blobs. Those blobs have a thin shell of cooled, solidified crust, but are still liquid or soft inside. When those blobs fall to rest on the bottom of the body of water in which they are formed, they tend to conforms to the bottom assuming the appearance of a random pile of pillows made of basalt.

References to that ancient ocean are common in discussions of theories of spontaneous generation in the search for the origin of life. However, it appears to be one of those things commonly accepted, but to which no particular scientist is given credit for discovering. At least, no big deal has been made of giving anyone credit for that discovery.

Bible attitude on earth covered by ocean

But the bible does make a big deal of it. It is mentioned in several creation accounts. It is the subject of the second verse of the bible— in the introduction to creation. That is because the emergence of the continents is important, and the ocean is what the continents emerged from. In the bible a statement of that fact is put up front. It is in the second verse of the first chapter of the first book of the bible:

[1a] *In the beginning*
[1b] *God*
[1c] *created*
[1d] *the sky*
[1e] *and the land.*
[2a] *And the land was not formed,*
> The original language word used here for land
> > referred to the continents.
> The beginning of the land was another
> > separate specific event
> at the emergence of the continents.

[2b] *there was none;*
> No continents existed, Ocean Covered
> > Planet Earth.

[2c] *and darkness*

> ^{2d} *covered the surface of the ocean,*
> ^{2e} *and a powerful wind [the wind of God] blew*
> ^{2f} *across the surface of the waters.* "

> Moses, c. 1445 BC, *Introduction to Creation II,*
> *A Biblical Creation Account:* Genesis 1:1,2

The bible published it first—there was a time when the planet earth was completely covered with ocean before there were any continents.

Science on How the Oceans were Filled:

The phenomenon of common acceptance (science by consensus) is encountered when one asks where the water of the ocean came from. Many theories have been put forth and rebutted. For many years, apparently it had just been assumed that the water to fill the oceans came from the interior of the earth. The Old Scientist has found no one who has been credited with that discovery.

In 1951, William Rubey[27] published a classic work demonstrating the lack of any mechanism for filling the ocean. However, by common acceptance, (since just about anyone can look out and see the ocean,) modern science has accepted the idea that the oceans of the earth are filled. No one has been credited with that discovery

But then science was rudely awakened.

To Louis Frank and John Sigwarth goes the credit for awakening the scientific community to the idea that the water of our oceans came from extra terrestrial source or sources.

Quite by accident, while studying another problem altogether, Louis Frank and John Sigwarth made a possible discovery and published a hypothesis that upset the scientific community. It was said of their hypothesis, that if it were true, then half the physical science books would have to be destroyed.

The essence of their new hypothesis is that the water that filled the oceans came from an extra-terrestrial source.

---

[27] **Rubey**, *William W. 1951. Geologic history of sea water: an attempt to state the problem. Geological Society of America Bulletin 62:1111-1148.*

RUBEY W.W., 1975. Geologic history of sea water; an attempt to state the problem. In: Kitano Y., (ed), Geochemistry of Water. Dowden, Hutchinson & Ross Inc., Stroudsburg, Pennsylvania.

The publication of that hypothesis was met with extreme disbelief in the scientific community. The disbelief was not so much with the hypothesis, but with the idea that another scientist would have the insolence to suggest that what scientists had previously believed was wrong. It had the effect of waking up the scientific community to a new idea. They had not considered the possibility of our oceans being filled with water from outer space.

At first, that hypothesis—more accurately, the interpretive theory of small comets bombarding the earth with a huge quantity of water daily—was met with extreme resistance by the scientific community. At first, anyone who sided with the discoverers was ridiculed by the resistance. Eventually, after a few years and no other satisfactory interpretation of the accumulating evidence, some began to accept the idea that the water of the oceans may have extra terrestrial origin.

In any case, whether the small comets are ever captured and proven to the satisfaction of doubters, or some other explanation is discovered, the attitude of modern science toward the water in the solar system, the universe, even the hydrologic cycle on earth has dramatically changed since the publication of the Small Comet Theory.

A brief history of the attitude of the scientific community is in order:

In 1951, William W. Rubey[28], a well-respected geologist, published his classic paper, *Geologic history of seawater: an attempt to state the problem.* summarizing the knowledge of science concerning the water of the earth, and the problems with the then current state of the knowledge. At that time, the water in the oceans of the earth was assumed to be the product of the out gassing of the rocks within the earth. In other words, ocean water came from juvenile water from the interior of the earth, water never before having been present on the surface. The problem was the water that could have possibly been from that source fell short by many orders of magnitude of being able to fill the oceans, even is the earth were considered to be a terrarium, never loosing any water to outer space.

---

[28] *Rubey, William W. 1951. Geologic history of sea water: an attempt to state the problem. Geological Society of America Bulletin 62:1111-1148.*

In 1983, just three years before Frank and Sigwarth published their theory of small comets, Shiklomanov and Sokolov[29] published a paper stating the earth was essentially a terrarium, neither gaining nor loosing water neither from nor to outer space.

By 1985, even the small and insufficient source of water coming from out gassing of the mantle being the source of water for the oceans was in doubt. Van Andel[30], in his, *New views on an old planet - A history of global change*, suggested that subduction was carrying water to great depths into the mantle at a faster rate than water is being returned to the surface by out gassing. Thus, the assumed source of water for the oceans was possibly in fact a loss, not a gain of water for the oceans.

By this time, the scientific community, even though most of the scientists did not realize it, were at a total loss for an explanation of the source of water in the oceans of the earth.

About that same time, quite by accident, evidence was discovered that indicated the possibility of the presence of a large quantity of water coming in from outer space in the form of giant snowballs about the size of a house. No one had observed these snowballs, but photographs of the earth's atmosphere had spots that appeared to be instrument noise and a scientist at the university of Iowa was given the task of solving the problem. However, the assumed instrument noise, was not correlated to any internal cause, rather, it seemed to be correlated with some unknown external cause. Thus the theory of Small Comets was born.

In 1986, Frank and Sigwarth[31] introduced the theory of small comets. This theory postulated the idea that small comets were bringing water into the earth's upper atmosphere where it is captured and added to the water of the oceans of the world. Further, the evidence indicated that the inflow of water was sufficient to fill the oceans and have a lot left over for loss to outer space, and to the

---

[29] *Shiklomanov, I.A., Sokolov, A.A. (1983). Methodological basis of world water balance investigation and computation. In New approaches in water balance computations. IAHS Publ. No. 148:77-90.*

[30] *Van Andel, T.H. 1995. New views on an old planet - A history of global change. Cambridge University Press*

[31] *Frank, L.A., Sigwarth, J.B., Craven, J.D. (1986). On the influx of small comets into the Earth's upper atmosphere. Geoph.Res.Lett.,13 :303-310.*

interior of the earth, an idea which had seemed apparent and bothersome to most scientists, but which they conveniently ignored because they did not have any explanation of a sufficient source for water to fill the ocean, especially, for any excess water that may be lost to outer space, or lost to subduction.

The Small Comet theory was met with a most violent reaction from the scientific community. In a classic demonstration of confirmatory bias toward the traditional beliefs of scientists, the theory was soundly criticized, and condemned. Further publication and research was strongly resisted.

However, in 1991, just five years later, Dr Vincent Kotwicki[32] published, *"Water in the Universe,"* a peer reviewed article in the *Hydrological Sciences Journal*. This article is a restatement of the state of knowledge of water in the universe published forty years earlier by Rubey when the earth was considered a terrarium. In this newer publication, the scope of hydrology on earth was proposed to be expanded to cover phenomena encountered on other celestial bodies. Kotwiki presented the state of the science at that time as considering the hydrologic cycle on earth as just an extension of an even greater hydrologic system in the solar system, and even extending to the open universe. (But hadn't that hydrologic cycle in outer space been described in the bible thousands of years earlier?)

In that publication, Kotwicki distanced himself from the infamous small comet theory. He dismissed it by stating that the small comet hypothesis had been disproved by Kerr[33] three years earlier. Even though that supposed disproof was based on the false assumption that it was obviously instrument noise, it served to isolate Kotwicki from accusations of influence by the small comet theory.

After eight more years pass, in 1999, David Deming[34] of the School of Geology and Geophysics at the University of Oklahoma in

---

[32] *Dr Vincent Kotwicki, (1991) Water in the Universe,* The Hydrological Sciences Journal, *J.Hydr.Sci.,36, 1,2/1991, pp49-66*

[33] *Kerr, R.A. (1989). Comets were a clerical error. Science.,241:352.*

[34] *Deming, David, 1999. On the Possible Influence of Extraterrestrial Volatiles on Earth's Climate and the Origin of the Oceans.* Palaeogeography, Palaeoclimatology, Palaeoecology *Volume 146, Issues 1-4, 15 February 1999, Pages 33-51. Abstract: http://www.sciencedirect.com/science?_ob=ArticleURL&_udi=B6V6R-3VX8YY5-3&_user=10&_coverDate=02%2F15%2F1999&_alid=566737579&_rdoc=1&_fmt=sum*

Norman, Oklahoma, published a classic paper in which he openly and seriously considers the small comets theory. No longer is an author obligated to distance himself from Frank and Sigwarth. In this paper, Deming states:

> "A consideration of observational and circumstantial evidence suggests that Earth may be subject to high influx rates (10(11)-10(12) kg/yr) of extraterrestrial-sourced volatile elements (carbon, hydrogen, oxygen, nitrogen) derived from comets or other primitive solar-system material. The total extraterrestrial influx rate may be four to five orders of magnitude greater than previously thought, large enough to account for today's total near-surface inventories of water and carbon."

In the intervening years since the announcement of the infamous theory of small comets, evidence has been collected which has been interpreted to both prove the existence of small comets, or to disprove the existence of small comets, depending on the bias of the scientist doing the interpreting.

All the while, modern science has been drifting toward considering the hydrologic system (water cycle) of the earth as an open system and acceptance of an extraterrestrial source for the water of the oceans.

In any event, whether the Small Comet Theory be true or just another temporary hypothesis on the way to the truth, to Frank and Sigwarth goes the credit for awakening the scientific community to the simple truth that the water of our oceans came from extra terrestrial source or sources.

This has been a tremendous change of attitude by modern science toward what was originally published thousands of years ago in the ancient scripture, yet unknown to modern science.

The Bible on How the Oceans were Filled:

Quite possibly their hypothesis is the key to the description of the origin of the oceans as found in the ancient scripture of the bible

mary&_orig=search&_cdi=5821&_sort=d&_docanchor=&view=c&_ct=1&_acct=C000 050221&_version=1&_urlVersion=0&_userid=10&md5=fea562ad65f6b720bb4d639da8 c28209

where Moses describes the mechanism where the atmosphere was formed and functioned to capture extra terrestrial water from outer space, and collect it below the atmosphere to fill the oceans

> *And God said, "let there be an atmosphere between the waters, and let it separate (set apart) water (onto the surface of the earth) from water [in outer space.] And God made this atmosphere and gathered (set apart) the waters [of the oceans] from the water above the atmosphere [in outer space.] And it really happened. And God called the atmosphere sky. And God said, Let the waters under the sky be gathered together unto one place, and let the dry [land] appear: and it was so. And God called the dry [land] continents; and the gathering together of the waters called he oceans:*

Moses, c. 1445 BC, On The Origin of the Atmosphere, The Origin of the Oceans, and The Emergence of the Continents. (from)
The Seven-Day Creation Account:
A Biblical Creation Account: Genesis 1:6-10

In his 1445 BC *One-Day Creation Account* Moses adds more detail of where the early water on the surface of the earth came from. Before there was an atmosphere, hydrologic cycle, or even an ocean, long before there was any human, the original water came from internal to the earth.

> [5c] *[At a time when] the LORD God had not yet made the rain on the earth* (hydrologic cycle),
> [5d] *and there existed no human to work the ground,*
> [6a] *vapor spewed up from the earth,*
> [6b] *and caused water*
> [6c] *on all the panorama of the ground* (whole surface.)

Moses, c. 1445 BC,
One-Day Creation Account,
A Biblical Creation Account: Genesis 2:4-7

Maybe that is where modern science got the theory that the water to fill the oceans came from juvenile water vapor out gassing from the interior of the earth. If so, they would not admit it. Even so, it was published in the bible thousands of years before modern science dreamed it up.

So, both mechanisms are published in the bible—juvenile water from the interior of the earth, and water from outer space both contributed to the water of the oceans. And thus the oceans were filled, by the combination of first juvenile water from the interior of the earth, then later by addition of water from extraterrestrial sources. This too, modern science will eventually realize. Modern science may even eventually hypothesize that the juvenile water and other gasses from the interior of the earth formed the first atmosphere, which then aided in capturing the water from the extraterrestrial sources.

In about 1015 BC, King David's, *Chronological Order of Creation* account, not only stated the fact that there was a time when the entire planet was covered with ocean, he even placed the filling of the oceans in the proper chronology position—after the laying of the mantle—and before the emergence of the continents.

> *⁵[Who] laid the foundations of the continents, (liquid*
> *   mantle)*
> *[that] it (continents) should not be removed for ever.*
> **⁶Thou coveredst it (mantle) with the ocean as [with] a**
> **   garment:**
> **the waters stood above the mountains.**
> *⁷At thy rebuke they (waters) fled;*
> *   at the voice of thy thunder they hasted away.*
> *⁸The mountains go up, the valleys go down*
> *   unto the place which thou hast founded for them (where*
> *   they are in equilibrium).*
> *⁹Thou hast set a bound that they (the oceans) may not pass*
> *   over;*
> *   that they (the oceans) turn not again to cover the earth.*
>
> David, c. 1015 BC,
> *Chronological Order of Creation,*
> *A Biblical Creation Account:* Psalms 104

Sometime around 970 to 930 BC King Solomon published his *Eyewitness Account by Wisdom* version of the creation story as seen by a superhuman eyewitness. In that account, he also got it right

63

[27b]*When he set a compass upon the face of the depth:*
He planned the ocean: 1. Atmosphere, 2. Extraterrestrial
water, 3. Continents emerge.
[28a]*When he established the clouds above:*
He established the atmosphere.
[28b]*When he strengthened the fountains of the deep:*
He added extraterrestrial water to the Juvenile
water to fill the oceans.
[29a]*When he gave to the sea his decree, that the waters
should not pass his commandment:*
Continents emerged and waters of the sea no
longer covered the land.
[29b]*When he appointed the foundations of the earth:*
When he set the continents on their
foundations.
Wisdom, c. 970 – 930 BC, *Eyewitness Account by
Wisdom,*
A Biblical Creation Account: Proverbs 8:1,12,22-31

The bottom line is all the writers of the bible got it right when they published it thousands of years before modern science discovered those same details.

Eon 3: Eon of Preparation for Complex Life

Eon 3 began with the emergence of the continents in a catastrophic event—a mantle turnover with tidal waves, darkness, and mountain building. Inhabitable land first appeared above sea level.

Before that the planet earth was covered with ocean. There were no continents.

This Eon started with that catastrophic event. That event culminated in an era of equilibrium.

Two major developments occurred in this Eon.

One was the establishment of land masses for land based life forms to exist in a balanced ecology.

The other was the development of many cycles of nature that eventually became the basis for a balanced ecology.

The accomplishment of these two developments paved the way for the next Eon, the Eon of Complex Life Forms.

3.0. Eon of Preparation for Complex Life
3.1.   The Emergence of the Continents - *The Earth (Land)*
3.1.1..   Then catastrophic event(s) occurred (Mantle overturn event(s)?)
3.1.2..   Then Continents began to emerge
3.1.3..   Then Continents established to be above sea level
3.1.4..   Then Continents are surrounded by continuous ocean,
              not oceans surrounded by continuous continent
3.1.5..   Then As continents grew there were episodes of mountain building,
3.2.   Cycles of Nature Established a Basis for Enduring Ecology
3.2.1..   Lithologic Cycle - Periodic Mountain Building and Eroding
3.2.2..   Hydrologic Cycle - Evaporation/Condensation Water
              Transportation
3.2.3..   . Then Life begins on earth
3.2.4..   . Carbon Dioxide/Hydrocarbon Cycle - Food chain
3.2.5..   . Reproduction Cycle—Biogenesis
3.2.6..   Chronobiology Cycles—Chronobiology Rhythms
3.2.7..   Food chain (Carbon Dioxide/Hydrocarbon cycle) fully developed
3.2.8..   Mass Extinction/Sustained Ecology Cycle—
              Punctuated Equilibrium

Note: The above table is a listing of only the major events of this eon. Details will be included as this section unfolds.

66

Modern science has discovered that before the continents formed, the planet earth was completely covered with oceans. Water stood above any mountains that existed. The oceans were filled before the continents formed. Modern science has not yet completely settled on a mechanism as to why or how this happened, but is certain that is the order in which it was.

After the land masses of the continents emerged above sea level and remained there as a steady state, the cycles of nature that are the basis of enduring land based ecology developed.

The mechanism that keeps the continental landmasses gathered up and above sea level consists of periodic mountain building episodes interspersed with long eras of eroding away. The eroding away never got so far as to cause the continents to be worn down to sea level. When The Old Scientist was in college, the question was being debated: Why did the continents not erode down to sea level as was predicted by the weathering hypothesis of the time? Now, the consensus of modern science believes they have figured it out: Plate tectonics keep the continents above sea level. Yet, the debate is not dead. There are competing hypotheses to explain some of the yet unexplained.

After the emergence of the continents put dry ground above the reach of the ocean waves, the Hydrologic Cycle functioned to maintain water far inland to support life.

Then the Carbon Dioxide/Hydrocarbon Cycle—the food chain—developed to furnish energy for life. Photosynthesis became the basis to convert sunlight into fuel.

Biogenesis developed where each form of life reproduced itself.

The chronobiology—natural rhythms—of life became synchronized with the annual, circadian, and lunar cycles.

The food chain became fully developed.

Finally, the cycle of Punctuated Equilibrium—with periods of sustained ecology followed by mass extinction then renewal of life forms—renewed every successive ecology with ever changing flora and fauna.

3.1 The Emergence of the Continents

The continents have not always existed. Somewhere in the far past history of this planet, the continents formed. They emerged from below sea level in one or more catastrophic events. Possibly it was an event known to modern science as a mantle turnover, a major continental movement event, sometimes called a surface renewal event.

A few years ago, Scientists had never heard of this. The common assumption was that the landmasses called continents had existed from the beginning. They had endured through the filling of the oceans.

But the bible has made a big deal of this. In the second verse of the bible, the subject was introduced—the land was not formed:

"In the beginning God created the sky and the land. And
the land was not formed, *there was none; and darkness covered the surface of the ocean [floor], and a powerful wind [the wind of God] blew across the surface of the waters."*

Moses, c. 1445 BC,
Introduction to Creation II,
A Biblical Creation Account: Genesis 1: 2

And it the same verse goes on to talk about the ocean that did exist.

Psalms 104 clarifies it with more detail, pointing out that, at that phase of development, the mountains were under the ocean.

⁵[Who] laid the foundations of the continents, (liquid mantle) [that] it (continents) should not be removed for ever. ⁶Thou **coveredst it** *(mantle)* **with the ocean** *as [with] a garment:* **the waters stood above the mountains.** *⁷At thy rebuke they fled; at the voice of thy thunder they hasted away. ⁸The mountains go up, the valleys go down unto the place which thou hast founded for them* (where they are in equilibrium). *⁹Thou hast set a bound that they (the oceans) may not pass over; that they (the oceans) turn not again to cover the earth.*

David, c. 1015 BC,
Chronological Order of Creation,
A Biblical Creation Account: Psalms 104

And the bible goes on to talk about it in many places. In fact, the continents emerging from the planet covered ocean is the second most mentioned topic in all the biblical creation accounts. It is mentioned more times than any other topic, second only to the most mentioned claim that God is what caused it.

3.1.   The Emergence of the Continents - *The Earth (Land)*
3.1.1.   Then catastrophic event(s) occurred (Mantle overturn event(s)?)
3.1.1..1.   Sea level established relative to continents (Sea held back)
3.1.1..2.   Continents (land) emerged ("brake forth", "issued out")
3.1.1..3.   Storm Clouds
3.1.1..4.   Darkness (debris in atmosphere)
3.1.1..5.   Tectonic Activity: Rapid subduction (breaking of tectonic plate?)
3.1.2..   Then Continents began to emerge
3.1.3..   Then Continents established to be above sea level
3.1.4..   Then Continents were surrounded by continuous ocean,
                    not oceans surrounded by continuous continent.
3.1.5..   Then As continents grew there were episodes of mountain building,
3.1.5..1   Mountains rose supported by floating on molten mantle (Isostatic
                    Rebound)
3.1.5..2.   Valleys fell
3.1.5..3   Isostatic balance developed

The Continents Formed

As this planet earth developed, long after it had cooled and the worldwide ocean had developed, a thin layer of lighter rock material was distributed over the whole planet, as a layer deep under the ocean. The continents existed only as a thin layer of continental crust material under the ocean.

Somehow, this worldwide thin layer of rock material became piled up into continents and emerged above sea level.

In the 1970's, a common question in college geology classes was concerning possible mechanisms which keep the continents above sea level. No one asked the question how did they get above sea level in the first place.

And the Bible Makes a Big Deal of it.

Even though science, and many man made religions, have long believed the continents existed forever, thousands of years ago, somewhere, it was known that the continents did not exist in the early history of this planet. The bible published the fact that the planet earth at one time had no continents above sea level, then, somewhat later; the continents emerged in a catastrophic event. All this happened in the far past, before human life existed on land.

The one who planned, engineered, and constructed the universe, yes, the creator, knew all along. And that intelligent being, yes, God, had it published. Apparently, that was for the purpose of letting people in modern times; know the God of the bible is the one who created heaven and earth. God not only published the claim to have created Heaven and Earth, but God gave detail not known by humans until the independent discovery of the same facts by modern science. In addition, He blinded theologians to that which he had published until after the separation of science and religion and the independent discovery of those same facts by modern science. That factual set of details is so accurate and so massive; that any scientist will have great difficulty honestly denying the claim that the bible published the facts thousands of years before modern science independently discovered the same facts.

In circa. 1445 BC, Moses, under the guidance of the creator of the universe, published these facts concerning the early history of this planet. His writings have survived these thousands of years as the religious writings of the one true religion established by the one true God who created the heavens and the earth, and all life therein.

Since science did not know the continents have not existed forever, theologians have had trouble interpreting that passage of ancient religious writings. They assumed that writing meant the world was just a chaotic jumble. Nothing could be further from the truth, neither of the true facts, nor of what that passage of scripture actually means. In fact, due to misunderstanding and misinterpretation, that whole section of scripture has been a black eye to theologians among scientists since the great separation of science and religion.

> *And the continents were not formed, there were none. And darkness covered the surface of the ocean. And the wind of God was blowing across the surface of the ocean."*
>
> Moses, c.1445 BC., The Holy Scripture,
> Genesis 1:2.

In circa. 1015 BC, King David, inspired by the creator, placed the same information into the ancient scripture. However, the information David placed into Holy scripture did not stop with the wind blowing. It went on to describe the catastrophic event that resulted in the emergence of the continents. As a bonus, it also described the principle of isostacy, discovered by modern science only a little over a hundred years ago.

> *[Who] laid the foundations of the continents, [that] it should not be removed for ever. Thou coveredst it with the ocean as [with] a garment: the waters stood above the mountains.*
>
> *At thy rebuke they fled; at the voice of thy thunder they hasted away. The mountains go up, the valleys go down unto the place which thou hast founded for them. Thou hast set a bound that they may not pass over; that they turn not again to cover the continents.*
>
> David, King of Israel, c.1015 BC, The Holy Scripture, Psalms 104:5-9.

Long before either Moses or David lived, Job quotes God himself on the same subject. After God mentions the planning, designing, and construction of the mantle as a foundation to support the continents, God goes on to describe the catastrophic event that resulted in the emergence of the continents from below sea level, indicating God's ancient knowledge that there was a time before which continents stood above sea level.

> *"Where wast thou when I laid the foundations of the continents? Declare, if thou hast understanding. Who hath laid the measures thereof, if thou knowest? Or who hath stretched the line upon it? Whereupon are the foundations thereof fastened? Or who laid the corner stone thereof; when the morning stars sang together, and all the sons of God shouted for joy?"*
>
> *"Or [who] shut up the sea with doors, when dry land brake forth, [as if] it had issued out of the womb? When I made the cloud the garment thereof, and thick darkness a swaddlingband for it, And brake up for it my decreed [place], and set bars and doors,*

And said, Hitherto shalt thou come, but no further:
and here shall thy proud waves be stayed?"
Job, Circa <1500 BC, The Holy Scripture, Job
38:4-11.

From the description of the darkness of the cloud in the same context, this appears to be the same catastrophic event described by Moses in Genesis 1:2.

Or, as referred to by Peter, nearly two thousand years ago, any of these passages may be referring to one of many repetitions of similar catastrophic events caused by the same mechanism.

> *"...and saying, "Where is the promise of His coming? For {ever} since the fathers fell asleep, all continues just as it was from the beginning of creation." For when they maintain this, it escapes their notice that by the word of God {the} heavens existed long ago and {the} continents were formed out of water and by water, through which the world at that time was destroyed, being flooded with water."*
> Peter, Circa 64-68 AD, The Holy Scripture, II
> Peter 3:4-6.

In any event, in these scripture passages, there were times in the past history of the planet we live on where there was water worldwide, covering the mountains [continents], and then there were times when the water of the oceans was prevented from inundating the continents.

Isostacy:

Another detail recorded concerning the emergence of the continents includes the mention of a principle undiscovered by modern science until about a hundred years ago—the principle of isostacy.

The bible does not explain the principle of isostacy to be where the depth of the mountain root determines the height of a mountain range. It does not explain that it is a geologic term referring to the equilibrium in the outer layer of the earth's crust where mountains and valleys are supported by the molten mantle at a height depending on their density and depth of root. The bible simply states that it happened.

72

> *"Mountains rose and valleys sank to the levels you decreed"*
>
> King David, *'The Bible"* ca. 1415 BC, NLT

It is interesting to note that when the bible was first translated into English, that particular passage was mistranslated because the translators could not imagine a principle of isostacy when it had not yet been conceived by modern science. No one believed mountains could actually rise and valleys sink to a predetermined level. Later, after that particular principle of science became accepted to be fact, subsequent translations dared to translate it properly.

A modern day science text book, in recounting the same event, might state it something like the following: After a catastrophic mantle overturn event, the mountains and valleys rebound to a height that is in equilibrium according to the principle of isostacy. The height of a mountain range and depth of a wide valley are determined by the amount of continental material beneath each. That amount of material, which is floating in the denser mantle below, lifts the mountain or valley to the point of equilibrium with the greater amount below the surface, similar to a floating iceberg,

The principle of isostacy is where the lighter rock material of the continents is floating in the denser mantle rock material below. That amount of material, which is floating in the denser mantle below, lifts the mountain or valley to the point of equilibrium with the greater amount below the surface. The more of that lesser dense continental rock material that is piled up, the deeper it is forced into the mantle below and thus exerts more upward pressure supporting the taller mountains.

Credit for discovering the principle of isostacy has been given to Clarence Edward Dutton[35] in about 1904 AD, over thirty-three hundred years after it was published by King David in about 1415 BC.

In this example—in a chronologically ordered list of scientific events—the bible simply states that it happened on a particular occasion. After mentioning the catastrophic emergence of the

---

[35] Dutton, Clarence Edward." *Earthquakes in the Light of the New Seismology,"* 1904.

continents, the bible specifically mentions that the mountains rose and valleys sank to the proper level. After that, the bible goes on with the list describing the hydrologic cycle and how it waters those mountains.

This pattern of casual mention of scientific principles is the key to understanding the wealth of scientific knowledge that is recorded in the bible.

It is assumed the idea did not come from discovery by modern science, because it was not discovered by modern science until thousands of years after it was recorded in the bible.

Doubting skeptics may have other theories as to the source of this information in the bible, but the bible itself claims this information came directly from the one true God, who, in his triune pre-incarnate (pre-universe) form of Intelligence, Wisdom, and Power, designed, engineered, and brought into existence, the universe itself. The bible claims the source of that ancient "scientific knowledge" within the bible came from the one who created the whole thing.

"He made the earth by his power, He engineered the world by his wisdom. He spread out the heavens by his intelligence."
Jeremiah, *'The Bible"* ca. 580 BC, (MOST)

3.2 Cycles of Nature

In this third Eon, *The Eon of Preparation for Complex Life*, is where the Cycles of Nature developed as the basis of enduring ecology.

3.2.   Cycles of Nature Established as the Basis of Enduring Ecology
3.2.0..   An era of equilibrium after continent and mountain building
3.2.1..   Lithologic Cycle Established - *To Maintain Continents*
3.2.2..   Hydrologic Cycle – *Waters the Continents*
3.2.3..   Preparation Complete for life Cycles to Develop – *Life Proliferates*
3.2.4..   Carbon Dioxide/Hydrocarbon Cycle - *Food chain*
3.2.5..   Reproduction Cycle—*Biogenesis*
3.2.6..   Chronobiology Cycles—*Chronobiology Rhythms*
3.2.7..   Food chain (Carbon Dioxide/Hydrocarbon cycle) fully developed
3.2.8..   Mass Extinction/Sustained Ecology Cycle-*Punctuated Equilibrium*

Eon 3 Part 2.0: An Era of Equilibrium
3.2.0..   An era of equilibrium after continent and mountain building

During this Era of equilibrium the cycles develop that provide for and maintain conditions where complex life forms can proliferate.

After describing the emergence of the continents, the *Chronological Order of Creation*, account in Psalms 104   simply states that a new equilibrium is established.

> *⁹Thou hast set a bound that they (the oceans) may not pass over; that they (the oceans) turn not again to cover the earth.*
>
> David, c. 1015 BC,
> *Chronological Order of Creation,*
> *A Biblical Creation Account:* Psalms 104

It then goes on to describe the new order of things. It goes on to mention many scientific principles working in that era that have a profound effect on ecology of the future eras.

Eon 3 Part 2.1: Lithologic Cycle
3.2.1..   Lithologic Cycle Established - *To Maintain Continents*
3.2.1..1.   Continent Maintaining Mechanism becomes Established
3.2.1..2.   Soil profile formed

Modern geology has long recognized that there is a cycle of periodic mountain building episodes followed by long periods of erosion wearing down the mountains. The Old Scientist has been unable to determine who first proposed, or described the evidence for such a cycle. It is just taught as fact without any fanfare about discovery. It has been proposed[36] that this cycle is driven by a mechanism of periodic surface renewal by rapid subduction (sinking) of the denser ocean floor into the less dense mantle material below. This periodic rapid sinking of the ocean floor gathers the lighter surface rocks which make up the continents, thus renewing the height of the mountains as the lighter material rebounds after such a surface renewal event.

When David mentioned the new equilibrium, he also mentioned the fact that a bound was set, a mechanism was put in action that maintained the continents to be above sea level continuously thereafter. Whatever that mechanism is finally to be settled upon by modern science, the fact that it even existed was not in the collection of stuff that modern science was aware of until recently—within the lifetime of The Old Scientist.

A possible candidate for that mechanism is known as the Lithologic Cycle, described in detail on pages 130 – 137 of this book in the section "A Possible Mechanism: The Lithologic Cycle."

---

[36] *Frederick, M. B., 1996, Origin of the Continents: An Introduction to the Theory of The Lithologic Cycle*, Max B. Frederick Publishing, 146 Laurel St., Central Point, Oregon 97502. *"Origin of the Continents"* may be purchased on the Internet by following the links from *www.EyewitnessToTheOrigins.com*

Eon 3 Part 2.2: Hydrologic Cycle

3.2.2..   Hydrologic Cycle – *Waters the Continents*
3.2.2..1.   According to pre-established laws of physics
3.2.2..2.   Hydrologic Cycle for purpose of sustaining life
3.2.2..3.   Sustained by water from the chambers. (incoming cosmic water)
3.2.2..4.   Hydrologic Cycle maintained for Future Time
3.2.2..5.   Earth (dry land, soil, continents) ready for life to begin,

No one should be shocked by the mention of the hydrologic cycle in the bible. It has long been known that it rains. Some religions imagine the rain is the tears of some goddess of virtue, but not the bible. Unlike other religions, somehow, the author of the bible knew rain came from the evaporation of ocean water and was a great unending cycle of nature.

> [20]*By his knowledge*
> *the depths* (oceans)
> *are broken up,* (evaporated)
> *and the clouds*
> *drop down the dew.*
> > Solomon, c. 930 BC, *Plan and Design,*
> > *A Biblical Creation Account:* Proverbs 3: 20.

Of course, it was thousands of years later when modern science first published the hydrologic cycle in 1788 when James Hutton published *Theory of Rain.*[37]

What is surprising is that the bible blatantly published the fact that the hydrologic cycle was sustained by water incoming from outer space. That is a concept which modern science has been dead set against until within the last couple of decades—and is only now coming to grips with. Modern science has long assumed all the water of our ocean had come from internal to the earth and the earth was neither gaining nor loosing water from or to outer space.

> [10]*He sendeth the springs into the valleys,*
> *[which] run among the hills.*

---

[37] *Theory of Rain, by James Hutton, in Transactions of the Royal Society of Edinburgh, 1788, vol. 1 , pp. 53-56.*

[11]They give drink to every beast of the field:
the wild asses quench their thirst.
[12]By them shall the fowls of the heaven have their habitation,
[which] sing among the branches.
[13]He watereth the hills from his chambers:
the earth is satisfied with the fruit of thy works.
[14]He causeth the grass to grow (for the cattle, and herb for
the service of man:)
that he may bring forth food out of the earth;

David, c. 1015 BC,
Chronological Order of Creation,
A Biblical Creation Account: Psalms 104

When David is describing the hydrologic cycle, he casually mentions that the source of the water for the hills is "his chambers." Those chambers just happen to be the other planets of our solar system. What modern science is in the process of discovering is the vast hydrologic cycle in the outer space of our solar system whereby water is transferred from one planet (chamber) to another. Thus, simply stated, the water supply on the planet earth is maintained at the cost of depletion of the water supply on the planet Mars. Modern science has not yet published that realization.

When Moses wrote of the source of the water for our oceans, he made it clear that it came from outer space.

> *"And God said, "let there be an atmosphere between the waters, and let it separate (set apart) water (onto the surface of the earth) from water [in outer space.] And God made this atmosphere and gathered (set apart) the waters [of the oceans] from the water above the atmosphere [in outer space.] And it really occurred. And God called the atmosphere sky. And God said, Let the waters under the sky be gathered together unto one place, and let the dry [land] appear: and it was so. And God called the dry [land] continents; and the gathering together of the waters called he oceans: "*

Moses, c. 1445 BC, On The Origin of the
Atmosphere, The Origin of the Oceans, and The

78

Emergence of the Continents. From the
Seven-Day Creation Account,
A Biblical Creation Account: Genesis 1:6-10

The scholars at the Great Library of Science at Alexandria, in about 250 BC who were the translators of the ancient Hebrew scriptures into the Greek LXX, were steeped in the culture and science of the day. They recognized the language of the bible that talked of water above the sky, as being a description of some scientific phenomena. However, the concept of a great snowstorm in space was not in their collection of knowledge. The modern science concept of dispersed water from outer space being the source of the water filling our oceans was not considered until the discovery in 1987 by Louis Frank at the University of Iowa, in the United States, many centuries later. Only God understood that concept at that time.

At that time, during the ancient Greek era, the science of the day included the false concept of solid crystalline spheres encircling the earth—which was the center of the universe—around which the entire geocentric (earth centered) universe rotated, with different parts of the universe connected to different concentric crystalline spheres rotating independently. Therefore, it was logical that their interpretation of the bible description of water above the sky would be a vast body of liquid water being held up by some solid, crystalline "firmament." Thus, the concept of such a firmament was introduced into the bible at the time of that translation. It is not in the inspired original language of the bible. God did not make that mistake, learned scholars did. Miraculously, God knew the simple truth.

Earth Ready for Life to Begin—Green things Begin to Grow

It is interesting to note that both David and Moses, after mentioning the Hydrologic Cycle, turn to the mention of green vegetation as the basis of the food chain.

David starts with saying the earth is satisfied with the fruit of God's works indicating the earth is ready for life to begin.

The earth is satisfied with the fruit of thy works.

> *[14]He causeth the grass to grow (for the cattle, and herb for*
> *the service of man:)*
> *that he may bring forth food out of the earth;*
> > David, c. 1015 BC, *Chronological Order of Creation,*
> > *A Biblical Creation Account:* Psalms 104

Moses simply states that God saw that it was good, indicating an intermediate stage of completeness.

[1:10c] *and God saw that [it was] good.*
> Indication of a completeness of this phase of creation.
> > Moses, c. 1445 BC,
> > *Seven-Day Creation Account,*
> > *A Biblical Creation Account:* Genesis 1

Then he just jumps right in to the photosynthetic vegetation as the basis for the food chain:

Eon 3 Part 2.4: Carbon Dioxide/Hydrocarbon Cycle-The Food Chain

3.2.4.. Food chain—Carbon Dioxide/Hydrocarbon Cycle develops

3.2.4..1. Photosynthesis Develops for CO_2 /Hydrocarbon cycle

3.2.4..2. Photosynthesis becomes the basis of the food chain

3.2.4..3. The atmosphere became habitable by photosynthesis

3.2.4..4. Food Chain becomes fully developed

3.2.4..4.1 Ultimate purpose of food chain is for sustaining human life.

As David's *Chronological Order of Creation* describes it, once the hydrologic cycle satisfied the requirement for water for the land, it was ready for land-based ecology to begin to produce food. The basis of the food chain is photosynthesis. Photosynthesis is here referred to as "grass" because that is the closest word in the Hebrew language to describe the greenness of the chlorophyll in green plants that is the active ingredient in photosynthesis. Technically, what it does is convert sunlight into food for all other forms of life. And the last phrase indicates the food chain's ultimate purpose is to provide food for the service of man.

> *The earth is satisfied with the fruit of thy works.*
> *[14]He causeth the grass to grow (for the cattle, and herb for*
> *the service of man:)*
> *that he may bring forth food out of the earth;*

David, c. 1015 BC,
Chronological Order of Creation,
A Biblical Creation Account: Psalms 104

Moses, (1445 BC) after the emergence of the dry land and the gathering together of the waters, indicated an intermediate stage of completeness by just jumping right in to the development of photosynthetic vegetation as the basis for the food chain:

1:11a *And God said, Let*

Again, it was all planned first.

1:11b *the earth sprout*

The earth was ready for life to begin.

1:11c *green vegetation, ...*

Photosynthesis would be necessary to convert light energy into chemical energy by combining carbon dioxide and water into hydrocarbon, thus establishing the basis for the food chain.

1:12a *And the earth sprouted*

Life developed on earth with photosynthesis providing the energy.

1:12b *green vegetation, ...*

Photosynthesis (Carbon Dioxide/Hydrocarbon cycle) developed.

1:29a *And God said, Behold, I have given you every herb ...; to you it shall be for meat.*

1:30a *And to every beast of the earth, and to every fowl of the air, and to every thing that creepeth upon the earth, wherein [there is] life, [I have given] every green herb for meat: and it was so.*

Photosynthesis—the process that uses sunlight to produce hydrocarbon from Carbon Dioxide and Water—is the basis of the food chain for all forms of life.

Moses, c. 1445 BC, *Seven-Day Creation Account,*
A Biblical Creation Account: Genesis 1

It is at this point in the chronology—after the beginning of photosynthesis and before animal life becomes abundant—that the atmosphere becomes habitable for modern animal life. That is due to photosynthesis producing oxygen. Of course, Carbon Dioxide is also essential for photosynthesis to produce oxygen. It is at this point in the chronology that the *Eyewitness Account by Wisdom* makes the statement:

[31a] *Rejoicing in the habitable part of his earth;*

Wisdom, c. 970 – 930 BC,
Eyewitness Account by Wisdom,
A Biblical Creation Account: Proverbs 8:1,12,22-31

Again, of course, this is all verified by the discoveries of modern science thousands of years after being published in ancient scripture.

The Discovery of Photosynthesis—by Modern Science:

Although the online encyclopedia, *Wikipedia* is less than reliable for facts, the following excerpts from *Wikipedia* demonstrate the significance of the early publication of the principle of photosynthesis as the basis of the food chain.

The following is found under the heading: Sunlight:[38]

> The existence of nearly all life on earth is fueled by light from the sun. Most autotrophs, such as plants, use the energy of sunlight to turn air into simple sugars—a process known as photosynthesis. These sugars are then used as building blocks and in other synthetic pathways which allow the organism to grow.

The following is found under the heading: Photosynthesis:[39]

> Although some of the steps in photosynthesis are still not completely understood, the overall photosynthetic equation has been known since the 1800s.
>
> Jan van Helmont began the research of the process in the mid-1600s when he carefully measured the mass of the soil used by a plant and the mass of the plant as it grew. After noticing that the soil mass changed very little, he hypothesized that the mass of the growing plant must come from the water, the only substance he added to the potted plant. His hypothesis was partially accurate - much of the gained mass also comes from carbon dioxide as well as water. However, this was a

---

[38] *Sunlight. (2007, March 9). In Wikipedia, The Free Encyclopedia. Retrieved 17:24, March 21, 2007, from http://en.wikipedia.org/w/index.php?title=Sunlight&oldid=113742523*

[39] *Photosynthesis. (2007, March 21). In Wikipedia, The Free Encyclopedia. Retrieved 17:09, March 21, 2007, from http://en.wikipedia.org/w/index.php?title=Photosynthesis&oldid=116801721*

---

signaling point to the idea that the bulk of a plant's biomass comes from the inputs of photosynthesis, not the soil itself.

Joseph Priestley, a chemist and minister, discovered that when he isolated a volume of air under an inverted jar, and burned a candle in it, the candle would burn out very quickly, much before it ran out of wax. He further discovered that a mouse could similarly "injure" air. He then showed that the air that had been "injured" by the candle and the mouse could be restored by a plant.

In 1778, Jan Ingenhousz, court physician to the Austrian Empress, repeated Priestley's experiments. He discovered that it was the influence of sun and light on the plant that could cause it to rescue a mouse in a matter of hours.

In 1796, Jean Senebier, a Swiss pastor, botanist, and naturalist who demonstrated that green plants consume carbon dioxide and release oxygen under the influence of light. Soon afterwards, Nicolas-Théodore de Saussure showed that the increase in mass of the plant as it grows could not be due only to uptake of CO_2, but also to the incorporation of water. Thus the basic reaction by which photosynthesis is used to produce food (such as glucose) was outlined.

Modern scientists built on the foundation of knowledge from those scientists centuries ago and were able to discover many things.

The significance of the early publication in the bible dating back to 1015 BC, and 1445 BC is emphasized by the big deal that is made in the final sentence of the above excerpt. In this online encyclopedia an issue is made of the ancientness of the discovery by calling it "centuries old." Yet, that "centuries old." discovery by modern science is only slightly over two centuries old. In comparison, the publication by God in the bible is over fifteen times as old.

1600-1800 AD Discovery and Publication by modern science is at most only a little more than four centuries ago.

1015 BC (publication by David) is over thirty centuries ago.

1445 BC (publication by Moses) is over thirty four centuries ago.

Science of the food chain:

It has long been known that there is a food chain, that animals eat other animals and those animals eat grass. However as a quantitative science, the idea that the sole source of energy for animal life comes

83

from the sun through photosynthesis in plants is relatively new. Two pioneers in that science— Alfred J. Lotka and Vito Volterra—date publication of their major discoveries back to 1925 and 1926, within the past century. Prior to that the food chain was recognized as a source of food, but not recognized as so dependent on photosynthesis and sunlight. Yes, just as the bible published first, light is the basis of the food chain.

1:29a *And God said, Behold, I have given you every herb ...; to you it shall be for meat.*

1:30a *And to every beast of the earth, and to every fowl of the air, and to every thing that creepeth upon the earth, wherein [there is] life, [I have given] every green herb for meat: and it was so.*

Moses, c. 1445 BC, *Seven-Day Creation Account*,

A Biblical Creation Account: Genesis 1

Yes again, modern science has discovered the same thing the bible published thousands of years earlier—the sun, through photosynthesis, is the source of food to every beast of the earth, and to every fowl of the air, and to every thing that creepeth upon the earth wherein there is life.

Eon 3 Part 2.5: Reproductive Cycle—Biogenesis

Biogenesis is life reproducing itself after its own kind.

3.2.5..   Reproduction Cycle—Biogenesis
3.2.5..1. Life became abundant on earth
3.2.5..1.1. Plant biogenesis developed
3.2.5..1.2. Animal biogenesis developed

First Plant Biogenesis:

*And God said, Let the earth bring forth grass, the herb yielding seed, [and] the fruit tree yielding fruit **after his kind**, whose seed [is] in itself, upon the earth: and it was so.   And the earth brought forth grass, [and] herb yielding seed **after his kind**, and the tree yielding fruit, whose seed [was] in itself, **after his kind**: and God saw that [it was] good.*

Then Animal Biogenesis:

84

> *And God created great whales (sea monsters), and every living creature that moveth, which the waters brought forth abundantly, **after their kind**, and every winged fowl after his kind: and God saw that [it was] good. And God blessed them, saying, Be fruitful, and multiply, and fill the waters in the seas, and let fowl multiply in the earth.*

> *And God said, Let the earth bring forth the living creature **after his kind**, cattle, and creeping thing, and beast of the earth **after his kind**: and it was so. And God made the beast of the earth **after his kind**, and cattle **after their kind**, and every thing that creepeth upon the earth **after his kind**: and God saw that [it was] good.*

> Moses, c. 1445 BC, On Biogenesis. From the
> *Seven-Day Creation Account,*
> *A Biblical Creation Account:* Genesis 1

Biogenesis is the process of life forms producing other life forms, e.g. a spider lays eggs, which form into spiders.[40]

Abiogenesis is life arising from non-living matter.

Spontaneous generation is life arising spontaneously—without interference by an intelligent being—from non-living matter.

Creation of life is some form of intelligence setting up the conditions for abiogenesis to occur—*let the earth bring forth…*

The bible consistently teaches creation followed by biogenesis.

Science—both modern and ancient—insists that abiogenesis (without intelligence) is the origin of all life.

Human beings have directly observed neither abiogenesis nor creation. Both scientists and theologians have given up on observing abiogenesis directly. Theologians rest on observing what the bible says. Scientists continue in their attempts to prove creation—abiogenesis plus their intelligence—unaware that what they are

---

[40] *Biogenesis. (2007, March 9). In Wikipedia, The Free Encyclopedia. Retrieved 11:57, March 22, 2007, from*
  http://en.wikipedia.org/w/index.php?title=Biogenesis&oldid=113938047

attempting to prove is creation. So far, even with the best intelligence available, the success has been very limited. Thus, the probability of spontaneous generation—abiogenesis without intelligence—is rapidly diminishing.

There is no disagreement between science and the bible as to what happened and when it happened, the disagreement is as to whether or not there was some form of intelligence that caused it to happen. In a short summary account of all creation, Moses summed up the part about the creation—abiogenesis plus intelligence—of man thusly:

> *And the LORD God formed man [of] the dust of the ground, and breathed into his nostrils the breath of life; and man became a living soul.*

> Moses, c. 1445 BC,
> *One-Day Creation Account,*
> *A Biblical Creation Account:* Genesis 2:4-7

Modern science continues to insist that spontaneous generation is possible, and in fact happened and is the origin of all life.   The "official" (politically correct) position of modern science can be summed up thusly: "… life can arise from non-life under suitable circumstances, although these circumstances still remain unknown."[41]

Beyond the disagreement concerning the original origin of life, both the bible and modern science agree that biogenesis took over and is the rule. They also agree on when it occurred. But, the bible published it thousands of years ago. Later, modern science came to the same conclusion that after the first life was formed, biogenesis took over. It was in 1859—the same year Charles Darwin first published Origin of the Species—when Louis Pasteur finally, scientifically, disproved the then politically correct version of spontaneous generation. It was at that time—about one hundred fifty years ago—the same year Oregon became a state, that the current version of the theory of spontaneous generation—the theory of evolution—became the latest theory of spontaneous generation.

---

[41] *Biogenesis. (2007, March 9). In Wikipedia, The Free Encyclopedia. Retrieved 11:57, March 22, 2007, from*
  http://en.wikipedia.org/w/index.php?title=Biogenesis&oldid=113938047

Eon 3 Part 2.6: Chronobiology Cycles—Chronobiology Rhythms

Chronobiology is a field of science that examines periodic (cyclic) phenomena in living organisms.[42]   In the field of Chronobiology, modern science studies the naturally occurring time period cycles in various life forms.

3.2.6..   Chronobiology Cycles—Chronobiology Rhythms
3.2.6..1.   Lunar Cycle for Seasons
3.2.6..2.   Circadian/Annual Cycles
                  (Day/Night/Seasonal Patterns of Life Forms)

In   1729   Jean-Jacques   d'Ortous   de   Mairan   discovered[43] Chronobiology. He discovered the circadian rhythm of plants is internal to the plant, not simply a response to the stimulation by the daily light from the sun. He put a plant having a cycle of opening and closing its leaves every twenty-four hours—in synchronism to the sunlight—into a dark closet and discovered the opening and closing of the leaves continued on a twenty-four hour schedule—just as though the light of the sun were present. This is the first recorded observation of the independence of the biological rhythms of the light-dark cycle in living organisms. Thus, in 1729, modern science first discovered the principle first published in the bible that the sun was for synchronizing, not producing, the natural rhythms in life.

Prior to the discovery of chronobiology, modern science assumed the daily cycle of all life was controlled, or stimulated by the light of the sun. It was not recognized to be internal to the life form and only synchronized by the sun. They should have recognized that. It was published in the bible thousands of years before modern science discovered it.   But theologians misinterpreted that publication. Instead of recognizing chronobiology in the publication, they interpreted it to be a publication of the timing of creation of the physical sun and moon, rather than creation of the significance of the sun and moon to synchronize the chronobiology of life.

---

[42] *Chronobiology. (2007, March 14). In Wikipedia, The Free Encyclopedia. Retrieved 11:40, March 24, 2007, from*
   http://en.wikipedia.org/w/index.php?title=Chronobiology&oldid=115005720
[43] *De Mairan, JJ.d.O (1729) Observation botanique. Histoire de l'Academie Royale des Sciences, 35-36*

Since the first discovery of natural rhythms in life forms in 1729, modern science has observed natural rhythms—the annual migration of birds, the menstrual cycle of mammals, the daily cycles of sleep awake patterns, yes, even seven day biological rhythms in humans[44]— that operate independently of outside stimuli. The sun, and moon are only signs to synchronize that natural chronobiology. It is interesting that modern science has discovered naturally occurring seven-day cycles in the human body. This is interesting because other naturally occurring cycles are generally tied to things like the rotation of the earth, the sun, the moon, etc. But in the case of the seven-day cycle, there is no heavenly body for synchronization. Franz Halberg, who is considered the founder of modern chronobiology—who coined the term circadian—who founded the Chronobiology Laboratories at the University of Minnesota—, has done considerable scientific research into biological rhythms. He started his experiments in the 1940's. Curiously, he has recognized weekly—seven-day—cycles in the human—and other organisms—that are critical to health. The references[45] to this work are too numerous and complex for

---

[44] *Kiser, Kim, 2005,* Father Time *(on the work of the University of Minnesota's Franz Hallberg, M.D.,* Minnesota Medicine, *November 2005, Volume 88, Published monthly by the Minnesota medical Association, Retrieved 0600, March 24, 2007, from http://www.mnmed.org/publications/MNMed2005/November/face%20to%20face-Kiser.htm*
[45] *Halberg F, Reinberg A, Haus E, Ghata J, Siffre M:* Human biological rhythms during and after several months of isolation underground in natural caves., *Nat Speleol Soc Bull 1970, 32:89-115.*

Hillman DC, Siffre M, Milano G, Halberg F: Urinary about-84-hour (circasemiseptan) variations of a woman isolated in a cave and cosmic ray effects., *New Trends in Experimental and Clinical Psychiatry 1994, 10:173-178.*

Zerubavel E: The Seven Day Circle: The history and meaning of the week., *New York: Free Press 1985.*

Cornélissen G, Halberg F, Wendt HW, Bingham C, Sothern RB, Haus E, Kleitman E, Kleitman N, Revilla MA, Revilla M Jr, Breus TK, Pimenov K, Grigoriev AE, Mitish MD, Yatsyk GV, Syutkina EV: Resonance of about-weekly human heart rate rhythm with solar activity change., *Biologia (Bratislava) 1996, 51:749-756.*

Ulmer W, Cornélissen G, Halberg F: Physical chemistry and the biologic week in the perspective of chrono-oncology.

In vivo 1995, 9:363-374.

Cornélissen G, Engebretson M, Johnson D, Otsuka K, Burioka N, Posch J, Halberg F: The week, inherited in neonatal human twins, found also in geomagnetic pulsations in isolated Antarctica., *Biomedicine and Pharmacotherapy 2001, 55(Suppl 1):32-50.*

Thaela M-J, Cornélissen G, Halberg F, Rantzer D, Svendsen J, Pierzynowski SG: Extra-

elaboration here, but if you want to pursue it, it is quite interesting, including discovery of cycles found in humans who are socially isolated in a cave. This discovery of seven-day cycles in humans and the importance to health is significant in the study of science and the bible. In the absence of any external indication of such a cycle could an ancient man made religion have put such importance on the Sabbath commandment as to give the violation of that commandment the penalty of death, and to frame the creation account as an illustration for that commandment?

It has only recently become obvious to modern science, life forms are made with built in naturally occurring biological cycles. These cycles are now known to be independent of the naturally occurring stimuli—sun and moon cycles—that historically were assumed to be the cause. Therefore, just as the bible said thousands of years before it was discovered by modern science, the heavenly bodies, the lights in the sky, are signs to keep the cycles synchronized.

> [14]*And God said, Let there be lights in the expanse of the sky to divide the day from the night; and let them be for signs, and for seasons, and for days, and years:*
>
> [15]*And let them be for lights in the expanse of the sky to give light upon the land: and it was so.*
>
> [16]*And God made two great lights; the greater light to rule the day, and the lesser light to rule the night: [he made] the stars also.*
>
> [17]*And God set them in the expanse of the sky to give light upon the earth,*
>
> [18]*And to rule over the day and over the night, and to divide the light from the darkness:*
>
> Moses, c. 1445 BC, *Seven-Day Creation Account,*
> *A Biblical Creation Account:* Exodus 20:4-7,
> Genesis 1:3-2:3, Exodus 31:15-17

---

circadian (about-weekly, half-weekly and 8-hourly) variation in pancreatic secretion of piglets., *Proc. XXXIII Int. Cong. International Union of Physiological Sciences, St. Petersburg, Russia* abstract P041.36., *June 30–July 5 1997*

Díez-Noguera A, Cambras T, Cornélissen G, Halberg F: A biological week in the activity chronome of the weanling rat: a chrono-meta-analysis., *Abstract, 4° Convegno Nazionale, Società Italiana di Cronobiologia, Gubbio (Perugia), Italy* 81-82., *June 1–2 1996*

Eon 3 Part 2.7: Food Chain Fully developed

3.2.7..   Food chain (Carbon Dioxide/Hydrocarbon cycle) fully developed

With the completion of the basic cycles of life, the food chain was complete, fully developed, and functional and various species were all in synchronism with the signs in the sky.

> *[21]The young lions roar after their prey,*
> *and seek their meat from God.*
> *[22]The sun ariseth,*
> *they gather themselves together,*
> *and lay them down in their dens.*

David, c. 1015 BC,
Chronological Order of Creation,
A Biblical Creation Account: Psalms 104

In 1924, Lotka published *Elements of Mathematical Biology.* In 1926 Volterra published *Variazioni e fluttuazioni del numero d'individui in specie animali conviventi.* These two publications resulted in the Lotka-Volterra equations recognizing the dependence of the food chain on photosynthesis and sunlight.

Is it any wonder that light is emphasized in the up front creation account describing the origin of the food chain. Only the true God who planned and created it all would have known that detail that long ago.

Eon 3 Part 2.8: Cycle of Mass Extinction/Sustained Ecology –
Punctuated Equilibrium

3.2.8..   Mass Extinction/Sustained Ecology Cycle
3.2.8..1.   Ecology Destruction and Renewal Became Pattern of Existence
3.2.8..1.1.   with mass extinction
3.2.8..1.2   by inundation

Punctuated equilibrium is the terminology modern science has assigned to the fact that, in the past, there have been long periods of time when no new species have appeared. Then—as though suddenly—an episode of speciation occurs. New species arise, completely different than any species that existed before.

Punctuated equilibrium was first published by modern science in 1971 when Niles Eldredge and Stephen Jay Gould presented a paper, *Punctuated Equilibria: An Alternative to Phyletic Gradualism*, at the Annual Meeting of the Geological Society of America, proposing the new theory.[46]

> *"The history of succession of changes in lifeforms*
> *on the planet earth is not one of stately unfolding, but*
> *a story of homeostatic equilibria disturbed only*
> *"rarely" (i.e., rather often in the fullness of time) by*
> *rapid and episodic events of speciation."*
>
> Eldredge and Gould 1972
> paraphrased by The Old Scientist

Definitions:

Equilibrium: a long time of stable ecology where species do not change, where biogenesis is the rule.

Biogenesis: each species reproduces itself in kind from generation to generation.

Punctuated: a relative short time of change where new lifeforms—new species—arise.

---

[46] "Punctuated Equilibria: An Alternative to Phyletic Gradualism" (1972) pp 82-115 in "Models in Paleobiology", edited by Schopf, TJM Freeman, Cooper & Co, San Francisco, Eldredge, N. & Gould, S.J. Retrieved 10:00, March 22, 2007, from
http://www.blackwellpublishing.com/ridley/classictexts/eldredge.pdf

---

Equilibria: plural of Equilibrium, implying multiple repetitions of eras of equilibrium.

Speciation: The beginning of a new species. The bible calls it creation. The latest spontaneous generation theory of science calls it evolution.

Phyletic Gradualism: The tenant of Evolution theory where life forms gradually change from one species to another.

Example of punctuated equilibrium in the *Seven-Day Creation Account* (1445 BC):

First, Moses describes a speciation event when megafauna (monsters) species arose:

> *And God created great whales (sea monsters), and every living creature that moveth, which the waters brought forth abundantly, after their kind, and every winged fowl after his kind:*

Next, Moses describes an equilibrium era where ecology is dominated by megafauna:

> *and God saw that [it was] good. And God blessed them, saying, Be fruitful, and multiply, and fill the waters in the seas, and let fowl multiply in the earth.*

Next, Moses describes a speciation event of mammals species:

> *And God made the beast of the earth after his kind, and cattle after their kind, and every thing that creepeth upon the earth after his kind:*

Next, Moses mentions an equilibrium era where ecology is dominated by mammals:

> *and God saw that [it was] good.*

Next, Moses describes a speciation event for humans:

> *And God said, Let us make man in our image, after our likeness: and let them have dominion over the fish of the sea, and over the fowl of the air, and over the cattle, and over all the earth, and over every creeping thing that creepeth upon the earth. So God*

> *created man in his [own] image, in the image of God*
> *created he him; male and female created he them.*

Next, Moses describes an equilibrium era where humans dominate ecology:

> *And God blessed them, and God said unto them,*
> *Be fruitful, and multiply, and replenish the earth, and*
> *subdue it: and have dominion over the fish of the sea,*
> *and over the fowl of the air, and over every living thing*
> *that moveth upon the earth.*

If you call it Punctuated Equilibrium you are putting the emphasis on the long periods of stable ecology. If you call it Cycle of Mass Extinction you are putting the emphasis on the relatively short period of species extinction and renewal. In either case it is a phenomena that has been operating since the beginning of life on this planet. At the current time, modern science is in the throws of discovering the mechanisms that drive this cycle. The most popular politically correct notion—the latest science by consensus—currently is to blame the human species for destruction of other species by his impact on the ecology. But then, science by consensus did not exist until the human species began to dominate the earth. And, this cycle of Mass Extinction was well established long before the arrival of the human species.

An Example emphasizing Mass Extinction from the account, *Everlasting-to-Everlasting* (1446 BC):

First Moses mentions a mass extinction event

> 3a *Thou causeth mortal humanity to return from destruction;*
> *and commandeth:*

Then Moses mentions renewal after mass extinction:

> 3b *"Return, ye descendents of the human race."*

Then Moses mentions an era of equilibrium:

> 4a *For a thousand years in thy sight [are but] as yesterday*
> *when it is past, and [as] a watch in the night.*

Then Moses mentions another mass extinction:

> 5a *Thou carriest them away*

5b *with a flood; they go to sleep:*

Then Moses mentions renewal after mass extinction:

5c *in the morning [they are] like grass [which] groweth up.*

Then Moses mentions an era of equilibrium:

6a *In the morning it flourisheth,*

6b *and groweth up;*

Then Moses mentions another mass extinction:

6c *in the evening it is cut down, and withereth.*

7 *For we are consumed by thine anger, and by thy wrath are we troubled.*

Finally Moses mentions the coming judgment:

8 *Thou hast set our iniquities before thee, our secret [sins] in the light of thy countenance.*

Note: In this account Moses mentions three mass extinctions the human race has survived or will survive in the future—the destruction of the Garden of Eden, Noah's Flood, and the coming Great and Terrible Day of the Lord—then finally, he mentions the judgment.

Yet another example, Peter (c. 64-68 AD)

(Note: The tense of the verbs in the first couple of verses indicates a continuing repetition, not a one-time event):

5 *For this they willingly are ignorant of, that by the word of God the heavens were of old, and the earth standing out of the water and in the water:*

6 *Whereby the world that then was, being overflowed with water, perished:*

7 *But the heavens and the earth, which are now, by the same word are kept in store, reserved unto fire against the day of judgment and perdition of ungodly men.*

8 *But, beloved, be not ignorant of this one thing, that one day [is] with the Lord as a thousand years, and a thousand years as one day.*

9   *The Lord is not slack concerning his promise, as some men count slackness; but is longsuffering to us-ward, not willing that any should perish, but that all should come to repentance.*

10   *But the day of the Lord will come as a thief in the night; in the which the heavens shall pass away with a great noise, and the elements shall melt with fervent heat, the earth also and the works that are therein shall be burned up.*

11   *[Seeing] then [that] all these things shall be dissolved, what manner [of persons] ought ye to be in [all] holy conversation and godliness,*

12   *Looking for and hasting unto the coming of the day of God, wherein the heavens being on fire shall be dissolved, and the elements shall melt with fervent heat?*

Peter, c. 64-68 AD, *Eternity-to-Eternity,*
A Biblical Creation Account: 2 Peter 3

Only in recent times has modern science discovered this long operating cycle. Before that, it was well known that the bible had references to a great mass extinction in the past, and referred often to a future mass extinction called the end of the eon (world) or the coming Great and Terrible Day of the Lord. It was well known that the bible had references to long periods of quiescent ecology, between the end of the Garden of Eden and the flood of Noah, between the flood of Noah and the coming great and terrible day of the Lord. However, theologians traditionally—yet contrary to scripture— assumed that all references to past extinction referred to the one time event of Noah's flood. Little did theologians realize the bible chronicles several historical mass extinctions. The oversight was just a simple matter of confirmatory bias where theologians thought they knew all the bible had to say in the field of science. But then, this is not the first time theologians have been embarrassed by their confirmatory bias.

It was in 1971 when the phrase *Punctuated Equilibria* was first coined. Niles Eldredge and Stephen Jay Gould presented a paper, *Punctuated Equilibria: An Alternative to Phyletic Gradualism*, at the Annual Meeting of the Geological Society of America proposing the new theory.[47] The essence of the new theory is contained in the following paragraph from page 84.

> "(4) The history of life is more adequately represented by a picture of "punctuated equilibria" than by the notion of phyletic gradualism. The history of evolution is not one of stately unfolding, but a story of homeostatic equilibria disturbed only "rarely" (i.e., rather often in the fullness of time) by rapid and episodic events of speciation."

It is interesting to note that that same paper contains great elaboration on the fact that unbiased observation is a myth—all observation is colored by theory and expectation—and that theories act as "party lines" to force observation into preset channels, unrecognized by adherents who think they perceive an objective truth. The following paragraph from page 85 of that same classic document summarizes the problem of confirmatory bias among modern scientists.

> "(2) Theory does not develop as a simple and logical extension of observation; it does not arise merely from the patient accumulation of facts. Rather, we observe in order to test hypotheses and examine their consequences. Thus, Hanson (1970, pp. 22-23) writes: "Much recent philosophy of science has been dedicated to disclosing that a 'given' or a 'pure' observation language is a myth-eaten fabric of philosophical fiction.... In any observation statement the cloven hoofprint of theory can readily be detected.""

The irony in all this is the fact that in arguing for open mindedness for acceptance for their new ideas, they described the *"cloven*

---

[47] "Punctuated Equilibria: An Alternative to Phyletic Gradualism" (1972) pp 82-115 in "Models in Paleobiology", edited by Schopf, TJM Freeman, Cooper & Co, San Francisco, Eldredge, N. & Gould, S.J. Retrieved 10:00, March 22, 2007, from
http://www.blackwellpublishing.com/ridley/classictexts/eldredge.pdf

hoofprint of theory that is readily detected" in their own observations. The theory of evolution is even named in their statement:

> "The history of **evolution** is not one of stately unfolding, but a story of homeostatic equilibria disturbed only "rarely" (i.e., rather often in the fullness of time) by rapid and episodic events of speciation."

Where an *unbiased* statement could have been:

> "The history of **succession of changes in lifeforms on the planet earth** is not one of stately unfolding, but a story of homeostatic equilibria disturbed only "rarely" (i.e., rather often in the fullness of time) by rapid and episodic events of speciation."

Assuming this is a statement of their observations, the inclusion of the word, "evolution" is a statement made in error. They did not actually observe evolution—the changing of one species into another—they only saw that there was a succession of changes where one species replaced another. They did not actually see one species turn into another. They assumed—according to their theory and bias—that what they observed was in fact one species spontaneously changing sufficiently to become another species. Yet, what they in fact saw, was nothing other than one species becoming extinct and another filling that same or similar ecological niche.

Inundation?

The bible clearly states that inundation is a mechanism for mass extinction. Aside from the story of the flood of Noah, Peter clearly states a mechanism involved in mass extinction is inundation.

> 6   *Whereby the world that then was, being overflowed[48] with water, perished:*
>
> > Peter, c. 64-68 AD, *Eternity-to-Eternity,*
> > *A Biblical Creation Account:* 2 Peter 3

However, modern science—even though they recognize all the signs of massive flooding—has not accepted flooding as a mechanism

---

[48] *The tense of the original language verbs here indicate a repeating phenomenon, not just a one time event.*

involved in the episodic mass extinctions. Therefore, you will notice, in the listing of details of science found in the bible, The Old Scientist has indicated that this is a bible only point of detail by making the numerical indicator appear in non-bold italics versus non bold roman which would indicate science only, or bold roman which would indicate both science and the bible:

3.2.8..   Mass Extinction/Sustained Ecology Cycle—Punctuated Equilibrium
3.2.8..1.   Ecology Destruction and Renewal Became Pattern of Existence
3.2.8..1.1. with mass extinction
3.2.8..1.2 by inundation

Modern science is in the throws of discovering the actual mechanisms involved in mass extinction.

It appears quite evident that the great mass extinction that ended the rein of the megafauna—and opened ecological niches for mammals—involved the impact of a giant meteorite with the earth. Various theories abound as to what the impact caused that resulted in mass extinction. One theory—the cosmic night theory—is that debris in the atmosphere blocked the light of the sun sufficiently that the lack of light, and the cold, extinguished vast amounts of life. Other theories abound. However, none of them sufficiently explain the rapid burial of massive remains of megafauna sufficiently to allow fossilization before deterioration. A massive flood would provide a mechanism for both the mass extinction and rapid burial. However, modern science is reluctant to consider such a scenario for two reasons. The first reason is the lack of a mechanism to provide sufficient water for such a flood. The second reason is their commitment to the philosophy of uniformitarianism—and the resistance of the scientific community to anything that even remotely resembles an influence of biblical Catastrophism. Even so, as time progresses, there is more acceptance of catastrophic episodic event involvement in shaping the ecology of today.

Eon 4: Eon Of Complex Life Forms

Eon 4 began with the emergence of complex life forms after a catastrophic mass extinction of the more primitive life forms of the previous ecology.

The Eon of Complex Life Forms has a clearly established pattern of abundance of life then catastrophe causing major loss of species. Each time, somehow, ecology and life forms repopulate the earth. Each time, the life forms are different than they were before. It is as though the former had to be destroyed to make way for the replacement ecology. This is the opposite of survival of the fittest to go on and populate the earth. It is destruction of the dominant fittest so the next wave of species can go on to populate the earth. It is as though the champions in the former ecology are destroyed to make way for the underdogs to become the new champions. How could this be?

The bible describes many times in the past where ecology has been destroyed, and then God has renewed the ecology. Even the future as depicted in the bible will have a major catastrophic destruction event called "the great and terrible day of the Lord." After that is described a new heaven and earth –a future continuation of the pattern.

4.0. Eon of Complex Life Forms
4.1.   Peak of Habitability—Ecology of Global Warming
4.1.1..   Era of sustained ecology.
4.1.1..1.   The sea swarms with living creatures.
4.1.2..   Then Plants with seeds developed.
4.1.3..   Era of Megafauna (Monsters)
4.2.   Major Mass Extinction (Cretaceous/Cenozoic Boundary Event)
4.3.   Ecology Restored, less Global Warming, Rise of Mammals
4.3.1..   Restoration of ecology - *Renewest the face of the earth*
4.3.2..   Era of sustained (renewed) ecology.
4.3.2..1.   Mammals appear
4.3.2..2.   Forests and grasslands
4.3.2..3.   Human species appears on earth
4.4.   Another Mass Extinction (Miocene/Pliocene Boundary Event)
4.4.1..   End of Tropical Ecology - *End of Garden of Eden*
4.5.   Ecology Restored, Global Cooling
4.5.1..   Era of restored ecology (Long time passes.)

4.6.   Another Mass Extinction
                    (Pleistocene/Holocene Boundary Event)
4.6.1..1. Mass Extinction of Large Mammals
4.7.   Ecology Restored/Epoch of Secular Human History
4.8.   Era of restored ecology
                    Proliferation of civilization after near extinction
4.8.1.. Beginning times/Current Era Boundary
4.8.2.. Living in Present
4.8.3.. Living In Future
4.8.3..1. Time of Sustained Ecology.
4.9.   End of Era of Complex Life Forms Event

Note: The above table is a listing of only the major events of this eon. Details will be included as this section unfolds.

Eon 4 begins with the filling of the sea with complex living creatures,[49] and ends with the coming future "Great and Terrible Day of the Lord."[50] It encompasses the time of all complex life forms on this planet including all existence of humankind until the final mass extinction—at least until the extinction of humankind on this planet. This final event—at least for humankind—will usher in Eon 5, the Eon of Eternity Future.

According to the bible, this eon witnessed many mass extinctions and renewals of ecology, each time with new life forms. At first the sea swarmed with living creatures, then plants with seeds first

---

[49] **Moses c. 1445 BC.**, *"And God said, Let the waters bring forth abundantly the moving creature that hath life..." Gen 1:20 (KJV)*
David, c. 1015 BC., *"...this great and wide sea, wherein [are] things creeping innumerable, both small and great beasts." Psalms 104:25 (KJV)*

[50] **Joel, c. 835 BC.**, *"And I will shew wonders in the heavens and in the earth, blood, and fire, and pillars of smoke. The sun shall be turned into darkness, and the moon into blood, before the great and the terrible day of the LORD come." Joel 2:30,31, (KJV)*
Peter, c. 64-68 AD., *"But the heavens and the earth, which are now, by the same word are kept in store, reserved unto fire against the day of judgment and perdition of ungodly men."*
"But the day of the Lord will come as a thief in the night; in the which the heavens shall pass away with a great noise, and the elements shall melt with fervent heat, the earth also and the works that are therein shall be burned up". 2Peter 3:7,10 (KJV)
John, c. 90-96 AD., *"And sware by him that liveth for ever and ever, who created heaven, and the things that therein are, and the earth, and the things that therein are, and the sea, and the things which are therein, that there should be time no longer:" Revelation 10:6. (KJV)*

appeared, then megafauna developed in the age of the dinosaurs, the first birds appeared, then flowering plants appeared…

Modern science agrees.

According to scientists, the earlier era—what they call the Cambrian Period—saw an abundance of sea life in an ecology that came to an end at the end of that era in a catastrophic mass extinction of the more primitive life forms of that previous ecology. In its place a new ecology arose with much more complicated life forms—even megafauna[51] life forms arose.

The bible says that early in this new eon there was an era that was the pinnacle of ecology. That pinnacle of ecology was the high point in the ecological history of this planet. It was the era in which the earth hit the peak of Habitability. It was populated with great animals of which God was proud.[52]

Modern science has also described that era as having a global warming produced tropical climate that made the earth habitable from pole to pole.

---

[51] *Giant animals, monsters, yes, even dinosaurs.*

[52] ***God, < 1500 BC.****, "Consider the behemoth, (dinosaur) which I made…" Job 40:15 From that introduction, God goes on to describe the wonders of a dinosaur, then some great sea monster.*

4.1: Pinnacle of Ecology

4.1.   Peak of Habitability—Ecology of Global Warming
4.1.1..   Era of sustained ecology.
4.1.1..1.   The sea swarms with living creatures.
4.1.2..   Then Plants with seeds developed.
4.1.3..   Era of Megafauna (Monsters)
4.1.3..1.   Megafauna (dinosaurs) thrive
4.1.3..2.   Pinnacle of Nature/Creation
4.1.3..3.   Sea monsters (Leviathan) thrive
4.1.3..4.   First birds developed
4.1.3..5.   First Flowering (Seeding) Plants.
4.1.3..6.   Ecology continues

The bible describes the ancient, former ecology on this planet as an ecology of paradise.   The ecology of today does not compare in grandeur to that of the ancient past—neither by bible standards, nor discovery of modern science.   Modern science has discovered that in the ancient past tropical rainforests covered the earth from pole to pole.   The ecology of today is barren and desolate[53] compared to that of the ancient past.

Over thirty-five hundred years ago, God published a description of the magnificent ecology early in the ancient past.   In that publication the very words of God describe the age of the dinosaur: God describes the dinosaur as "the chief of the ways of God," and the sea monster as "a king over all the children of pride."   In this account, God describes the vegetation of that era as what we today would call a rainforest.   (Note: The traditional translations of the bible do not use

---

[53] *Let me get on my scientist's soap box for a moment: The bible has commissioned humanity to improve the ecology it finds itself in.   Remember the parable of the talents.   Each was expected to increase the value of the investment entrusted to them.   Those who increased the value were rewarded.   Those who conserved it as it was, were condemned.   So much for the conservation of the comparatively barren and desolate ecology we are trying to conserve rather than improve.   We are living on a dying planet and are doing nothing to improve it.   Modern culture views it is a virtue to preserve ecology as it is.*

We can restore the rainforests of yesteryear by recycling the vegetation lost to the ecology that is trapped in the coal beds.   The ecology of today is limited by starvation for the raw materials with which to rebuild the rainforests.   We have the technology, but are blinded to both the need and the process…

the word "dinosaur." That word was not coined by modern science until after the bible was translated into English.)

> *Behold now behemoth, which I also made; he eateth grass as an ox. Lo now, his strength [is] in his loins, and his force [is] in the navel of his belly. He moveth his tail like a cedar: the sinews of his stones are wrapped together. His bones [are as] strong pieces of brass; his bones [are] like bars of iron. He [is] the chief of the ways of God: he that made him can make his sword to approach [unto him]. Surely the mountains bring him forth food, where all the beasts of the field play. He lieth under the shady trees, in the covert of the reed, and fens. The shady trees cover him [with] their shadow; the willows of the brook compass him about. Behold, he drinketh up a river, [and] hasteth not: he trusteth that he can draw up Jordan into his mouth....*
>
> *Canst thou draw out leviathan with an hook? ....*
>
> *Upon earth there is not his like, who is made without fear. He beholdeth all high [things]: he [is] a king over all the children of pride.*

God, c. <1500 BC,
The Pinnacle of Ecology,
A Biblical Creation Account: Job 40:15-41:34

In 1015 BC, over three thousand years ago, king David published a description of the ecology of that earlier era:

> [24]*O LORD, how manifold are thy works!*
> *in wisdom hast thou made them all:*
> *the earth is full of thy riches.*
> [25]*[So is] this great and wide sea,*
> *wherein [are] things creeping innumerable,*
> *both small and great beasts.*
> [26]*There go the ships: [there is] that leviathan,*
> *[whom] thou hast made to play therein.*
> [27]*These wait all upon thee;*
> *that thou mayest give [them] their meat in due season.*
> [28]*[That] thou givest them they gather:*
> *thou openest thine hand, they are filled with good.*

David, c. 1015 BC, *Chronological Order of Creation,*
A Biblical Creation Account: Psalms 104

Again, in 1015 BC, king David publishes another account that mentions that era in a positive note. Right after his mention of the emergence of the continents, and before his description of the mass extinction that wiped out the ecology of the age of megafauna, the report says:

7bPraise Him] ye monsters,

David, c. 1015 BC,
Before and After Account,
A Biblical Creation Account: Psalms 148

Moses, in 1445 BC, almost thirty-five hundred years ago published a description of the flora and fauna of that age of the dinosaur:

11 *And God said, Let the earth sprout green vegetation, the herb yielding seed,   [and] the fruit tree yielding fruit after his kind, whose seed [is] in itself, upon the earth: and it was so.*
12 *And the earth sprouted green vegetation, herb yielding seed after his kind, and the tree yielding fruit, whose seed [was] in itself, after his kind: and God saw that [it was] good...*
20 *And God said, Living souls shall be roamers of the waters and flyers above the earth in the open atmosphere of the sky.*
21 *And God created wonderful monsters, and all the living souls that roam the waters, after their kind, and every winged flyer after his kind: and God saw that [it was] good.*

Moses, c. 1445 BC,
Seven-Day Creation Account,
A Biblical Creation Account: Genesis 1

Moses, again in 1445 BC, published a description of the attitude of God toward the ecology of the ancient past:

8a The LORD God planted a garden in the luxurious [ecology] of the ancient past;...

[9a] *Out of the ground the LORD God caused to grow*
every tree that is pleasing to the sight and good for food; ...
Moses, c. 1445 BC,
The Ecology of Paradise,
A Biblical Creation Account: Genesis 2

Today, children's books abound describing the dinosaur and the lush ecology that dominated the age of the dinosaur. Hardly a child exists that is not aware of this glorious ancient past.

Yet, before 1824, modern science did not realize that the age of the megafauna (dinosaur) ever existed. They were completely blinded to the accounts of it in the bible. No one had imagined what the bible had published thousands of years earlier.

Although it is now realized that the existence of dinosaur bones has been known for thousands of years, it is only recently that they have been recognized as dinosaur bones.

In 1676 the first dinosaur bone to be described by modern science was discovered at Cornwell near Oxford, England. Shortly thereafter, Robert Plot[54] published an accurate description of the bone, but did not recognize it as belonging to any unknown species. Instead it was suspected that it might be the thighbone of an elephant. In reliance on a misinterpretation of the science of the bible prevalent at the time, it was even speculated that it might be of a giant now extinct human. No speculation was made that it might be from the ancient age of the dinosaur. The name "dinosaur" was unheard of until Sir Richard Owen coined it in 1842[55]

It was not until 1824—less than a mere 200 years ago, and yet 35 years before Darwin first published the theory of evolution—that William Buckland published[56] the realization that fossilized dinosaur bones were of a giant lizard-like species of ancient megafauna.

It has only been within the past twenty or so years that modern science has extended the dinosaur fossil findings to every continent, including Antarctica.

---

[54] *Plot, Robert, (1677).* Natural History of Oxfordshire
[55] *Owen, R. (1842). "*Report on British Fossil Reptiles.*" Part II. Report of the British Association for the Advancement of Science, Plymouth, England.*
[56] *Buckland, W. (1824). "Notice on the* Megalosaurus *or great Fossil Lizard of Stonesfield."* Transactions of the Geological Society of London, *series 2, vol. 1: 390–396.*

Yet, over thirty-five hundred years ago, God accurately published a scientifically accurate description of a dinosaur in the book of Job.[57]

---

[57] *God, < 1500 BC., "Consider the behemoth, (dinosaur) which I made…" Job 40:15 From that introduction, God goes on to describe the wonders of a dinosaur, the ecology of that era, and some great sea monster.*

4.2: Major Mass Extinction—End Megafauna Ecology

4.2.   Major Mass Extinction (Cretaceous/Cenozoic Boundary Event)
4.2.1..   Major Mass extinction (Mantle turnover event with extinction)
4.2.1..1.   Tectonic Activity End Cretaceous
4.2.1..2.   Catastrophic event (great storm) (tidal waves)
4.2.1..3.   Darkness (debris in atmosphere)
4.2.1..4.   Breakup of continents (Mantle overturn event?)
4.2.1..5.   Great flood *(breath taken away)*
4.2.1..6.   Extinction (death/dust)
4.2.1..7.   Isostatic rebound (mountain building episode)

Immediately after his description of the first glorious ecology of Eon 4—The Age of the Dinosaur—the *Chronological Order of Creation* account describes a major mass extinction. Scientists will recognize that as the Cretaceous/Cenozoic (Kt) Boundary Event—the catastrophic mass extinction at the end of the age of the dinosaur.

The Eon of Complex Life Forms:
The Age of Monsters (Dinosaurs)
> [24]*O LORD, how manifold are thy works!*
>   *in wisdom hast thou made them all:*
>   *the earth is full of thy riches.*
> [25]*[So is] this great and wide sea,*
>   *wherein [are] things creeping innumerable,*
>   *both small and great beasts.*
> [26]*There go the ships: [there is] that leviathan,*
>   *[whom] thou hast made to play therein.*
> [27]*These wait all upon thee;*
>   *that thou mayest give [them] their meat in due season.*
> [28]*[That] thou givest them they gather:*
>   *thou openest thine hand, they are filled with good.*

Major Mass Extinction — Cretaceous/Cenozoic (Kt) Boundary Event
> [29]*Thou hidest thy face,*
>   *they are troubled:*
> *Thou takest away their breath,*
>   *they die, and return to their dust.*

David, c. 1015 BC,
Chronological Order of Creation,
A Biblical Creation Account: Psalms 104

Many theologians, in the context of a Young Earth, interpret this out of context as referring to the cycle of life within a single generation. But, in context, it is a description of the continuing pattern of the cycle of mass extinctions.

In another publication, King David gives more details of the mass extinction. Just after his mention of the monsters in praise, King David speaks of the catastrophic event at the end of the age of the monsters.

History of Existence From the Emergence of the Continents Forward in Time.

   [7a] *Praise the LORD from the earth,* ("from the beginning of dry land")

Era of Megafauna: Age of the Dinosaur

   [7b][*Praise Him] ye monsters,*

End of Era of Megafauna: Mass Extinction. A (Kt) Boundary Event

   [7c]*And all the oceans,* [8a]*fire and falling judgment, falling white and*
        toxic fume, tempestuous wind,   (meteorite collision event)
   [8b]*All following his instructions* (all within the laws of physics).
   [9a]*Mountains, and all hills;*

David, c. 1015 BC, *Before and After Account,*
A Biblical Creation Account: Psalms 148:

King David even included the detail about the episode of mountain building that correlates with the aftermath of this catastrophic event.

In this day and age of myriads of children's books telling the story of the age of the dinosaurs, and their demise in a catastrophic event where a meteorite struck the earth, causing them to go extinct, it is hard to imagine anyone not knowing there is a well established pattern of repeating mass extinctions.

It was published over three thousand years ago in the bible.

Yet, until 1824—less than two hundred years ago—people, including theologians and scientists alike, were blinded to the facts published so clearly in the bible. Due to interpretation of the science of the bible being so biased toward the science of the Greek era, they could not imagine such a thing as the age of the dinosaur.

In fact, it was not until 1980—when Luis Alvarez Published *Extraterrestrial causes for Cretaceous-Tertiary Extinctions*—that the end of the dinosaur age was even considered to be such a significant event. It was only about thirty–five years ago that they proposed the Giant Asteroid Theory for Catastrophic mass extinction. Luis Alvarez and his son Walter described the discovery of the impact of a giant meteorite at the Yucatan Peninsula, and how it gave a catastrophic explanation of the end of the age of the Dinosaurs.

It is hard to believe that it was not seen before—it is so clearly described in the bible.

Even the Seven-Day Creation Account inserts the closing of a day at the end of the age of the dinosaur before going on to describe the fauna of the next renewal of ecology.

> [11]   *And God said, Let   the earth sprout   green vegetation, the herb yielding seed,   [and] the fruit tree yielding fruit after his kind, whose seed [is] in itself, upon the earth: and it was so.*
>
> [12]   *And the earth sprouted green vegetation, herb yielding seed   after his kind, and the tree yielding fruit, whose seed [was] in itself, after his kind: and God saw that [it was] good.*
>
> [20] *And God said, Living souls shall be roamers of the waters and flyers above the earth in the open atmosphere of the sky.*
>
> [21] *And God created **wonderful monsters**, and all the living souls that roam the waters, after their kind, and every winged flyer after his kind: and God saw that [it was] good.*
>
> [23] ***And the evening and the morning were the fifth day***. (Bold emphasis added.)
> > Moses, c. 1445 BC, *Seven-Day Creation Account,*
> > *A Biblical Creation Account:* Genesis 1

4.3: Ecology Restored—Rise of Mammals

4.3.   Ecology Restored, less Global Warming, Rise of Mammals

4.3.1..   Restoration of ecology - *Renewest the face of the earth*

4.3.1..1. New life forms Cycle Of Extinction & Renewal
                    (Punctuated Equilibrium).

4.3.1..2. Spread over continents The far past of recent life.

4.3.2..   Era of sustained (renewed) ecology.

4.3.2..1. Mammals appear

4.3.2..2. Forests and grasslands

4.3.2..3. Human species appears on earth

4.3.2..4. Man became a living soul

4.3.2..5. Tropical ecology with mild global warming

4.3.2..6. Punctuated Equilibrium Continues

Eon 4, The Eon of Complex Life Forms, has an established pattern of stable ecology with abundance of life followed by some catastrophe causing major loss of species. Each time, somehow, ecology and life forms repopulate the earth. Each time, the life forms are different than they were before.

In Eon 4, Part 2, we saw vivid description of a major catastrophic event that destroyed the ecology of planet earth long before the flood of Noah.

Here, in Eon 4, Part 3 we will see that immediately after the mass extinction, a new ecology arose. Then there was a long era of sustained ecology. In this era, mammals arose and the human species appeared on the earth. The climate was sub-tropical with mild global warming

The bible says these new species were created by the intelligent power, *Elohiym*, whom theologians call the almighty God.

Modern science says they spontaneously arose without the help of any intelligent being.

In King David's third creation account, *Chronological Order of Creation*, he describes this renewal of ecology—a restoration of the grandeur of the former age with a new generation of species—a creation of a new suite of species to renew the ecology of the planet earth.

Ecology Restored.
30a Thou sendest forth thy spirit,

30b they are created:

30c And thou renewest the face of the earth.

> David, c. 1015 BC, *Chronological Order of Creation,*
> *A Biblical Creation Account:* Psalms 104

The chronology here is important. It proves the ultimate author of the bible knew the correct chronology long before modern science discovered it. In King David's account, this description comes immediately after he described a glorious ecology of Eon 4—The Age of the Dinosaur—then the catastrophic major mass extinction that ended the age of the dinosaur. This is an example of an episode of what modern scientists continue to refuse to recognize as an episode of creation, but will acknowledge as an episode of punctuated equilibrium.

The Eon of Complex Life Forms:
The Age of Monsters (Dinosaurs)

Note: In verse 26 the translation uses the word "ships," but it is obviously referring to living beasts as the context reveals.

24O LORD, how manifold are thy works!
in wisdom hast thou made them all:
the earth is full of thy riches.
25[So is] this great and wide sea,
wherein [are] things creeping innumerable,
both small and great beasts.
26There go the ships: [there is] that leviathan,
[whom] thou hast made to play therein.
27These wait all upon thee;
that thou mayest give [them] their meat in due season.
28[That] thou givest them they gather:
thou openest thine hand, they are filled with good.

Major Mass Extinction — Cretaceous/Cenozoic (Kt) Boundary Event
29Thou hidest thy face,
they are troubled:
Thou takest away their breath,
they die, and return to their dust.

Ecology Restored.
30a Thou sendest forth thy spirit,
30b they are created:
30c And thou renewest the face of the earth.

Era of Sustained Ecology.

[31a] *The glory of the LORD shall endure for ever:*
[31b] *the LORD shall rejoice in his works.*

David, c. 1015 BC, *Chronological Order of Creation,*
A Biblical Creation Account: Psalms 104

Further, in his fourth creation account, in his *Before and After Account,* King David again has the chronology correct, and gives details of the exact class of species that show up: Mammals.

History of Existence From the Emergence of the Continents Forward in Time.

[7a] *Praise the LORD from the earth,*

Era of Megafauna: Age of the Dinosaur

[7b] *Praise Him] ye monsters,*

End of Era of Megafauna: Mass Extinction

[7c] *And all the oceans,* *[8a]* *fire and falling judgment, falling white and toxic fume, tempestuous wind,*

[8b] *All following his instructions (laws of physics).*

Restoration of Ecology:
    Mountain Building Episode

[9a] *Mountains, and all hills;*

    Age of Mammals

[9b] *Fruit trees, and all cedars:*

[10a] *The living things, and cattle, animals that move, and birds that fly;*

David, c. 1015 BC, *Before and After Account,*
A Biblical Creation Account: Psalms 148

Other Creation accounts provide more detail.

The *Seven-Day Creation Account,* recorded by Moses in 1445 BC, puts it this way:

The Era of Mammals

[24a] *And God said, Let the land bring forth the living creature after his kind, cattle, and creeping thing, and beast of the land after his kind: and it was so.*

Age of Mammals, planned by God,
to replace age of megafauna.

[25a] *And God made the beast of the land after his kind, and cattle after their kind, and every thing that creepeth upon the land after his kind:*

Age of Mammals
replaces age of megafauna.

²⁵ᵇ *and God saw that [it was] good.*
Indication of a completeness of this phase of creation.
Chronologically matches an era of equilibrium in ecology.

Genetic Engineering Humankind

²⁶ᵃ *And God said, Let us make man in our image, after our*
likeness: and let them have dominion over the fish of the
sea, and over the fowl of the air, and over the cattle, and
over all the earth, and over every creeping thing that
creepeth upon the earth.
God planned the creation of human race
²⁷ᵃ *So God created man in his [own] image, in the image of God*
created he him; male and female created he them.
Creation of human race
Moses, c. 1445 BC, *Seven-Day Creation Account,*
A Biblical Creation Account: Genesis 1

The Ecology of Paradise Creation account also recorded by Moses
in 1445 BC puts it this way:

¹⁵ *Then the LORD God took the man and put him into the*
Ecology of Paradise to cultivate it and keep it.
¹⁶ *The LORD God commanded the man, saying, "From any*
tree of the garden you may eat freely;..."
²⁵ *And the man and his wife were both naked and were not*
ashamed.   Indication of a mild tropical climate.
Moses, c. 1445 BC, *The Ecology of Paradise,*
A Biblical Creation Account: Genesis 2

The *Eternity-to-Eternity* Creation account, recorded by Peter in c.
68 AD emphasizes the fact that the cycle of mass extinctions was a
repeating pattern. In the original Greek language, the verbs used in
the account are in the aorist tense. There is no equivalent tense in
English language, the concept of this verb is considered without
regard for past, present, or future time, indicating something that
happened and continues to happen.

In verse five, Peter describes the mechanism whereby the land of
the continents are overflowed with water as being the same physical

114

mechanism operating according to the laws of physics (Intelligence of God) that caused the continents to originally emerge from beneath the surface of the ocean. Then in verse six, Peter applies that same principle of physics to the fact that the continents have since that time been subject to repeated overflowing with water resulting in repeated destruction episodes. It is interesting to note that theologians translating this passage according to their bias toward only one episode of flooding of the continents—the flood of Noah—have ignored the repeating inundation and mass extinction pattern indicated by this scripture.

The idea of this being indicative of a repeating pattern is reinforced by the continuation on into verse seven of the pattern. Verse seven describes yet another future mass extinction.

> 5 *For this they willingly are ignorant of, that by the word of God the heavens were of old, and the earth (*continents*) standing out of the water and in the water:*
>
> 6 *Whereby the world that then was, being overflowed with water, perished:*
>
> 7 *But the heavens and the earth, which are now, by the same word*   (same laws of physics, i.e. by same mechanism,)
>
>   *are kept in store, awaiting for fire against the day of judgment*
>
>   *and destruction of irreverent humans.*
>
>   Future mass extinction.
>
>   Peter, c. 64-68 AD, *Eternity-to-Eternity,*
>
>   *A Biblical Creation Account:* 2 Peter 3

The irony in the antagonism between the bible and science is the fact that modern science has described more repetitions of this cycle of punctuated equilibrium[58] with more detail, than is explicitly

---

[58] *Major mass extinctions, renewal events, appearances of species, etc. discovered by modern science includes:*

PALEOZOIC

Cambrian: *Sustained Ecology Era*

Continents break up (Plate Tectonics) Major Mass Extinction

Ordovician: *Sustained Ecology Era. Sea life becomes abundant. Followed by another major mass extinction.*

Silurian: *New era of sustained ecology. Climate stabilizes into more habitable global warming,*

outlined in the bible, yet with each new class of species showing up in the same order as listed in the bible. In other words, the bible accounts leave out unimportant detail; yet contain enough detail to prove knowledge in advance of scientific discovery.

Melting ice. Rise in sea level First vascular plants. First fish with Jaws
Devonian: *Plants have Seeds (DNA) Biogenesis. Era ends with another mass extinction*
Carboniferous: First Amniota eggs. (common to turtles, lizards, birds, dinosaurs and
    mammals)
Permian: *Era ends with largest mass extinction recorded in the history of the earth.*
MESOZOIC:
Triassic: *Continents again reform into one supercontinent Pangaea.*
Era of sustained ecology. *Survivors of mass extinction spread and recolonized.*
Ends with another mass extinction.
Jurassic: *Sustained Ecology. Pinnacle of Nature. First Birds*
Large life forms form and thrive
Cretaceous:
First Flowering (Seeding) Plants
Break up of supercontinent Pangaea.
Era ends with mass extinction caused by meteorite crash into earth.
Mass Extinction (by continental flooding?)
Cosmic Night/Great Storm
Order and Ecology Restored after Mass Extinction
Continuation of cycle of Extinction & Renewal (Punctuated Equilibrium)
CENOZOIC: Paleogene
Stable Ecology with repeating, lesser magnitude, Mass Extinctions.
Age of mammals, flowering Plants, Insects...
Tropical Ecology with Global Warming.
CENOZOIC: Neogene
Man appeared on earth
Minor Mass Extinction with Tectonic volcanism
Ecology Restored, Yet colder, harsher habitat.
Ice ages
Human species flourishes.
Another Mass Extinction. (North American Large Animals.)

4.4: Mass Extinction—End of Garden of Eden

4.4.   Another Mass Extinction (Miocene/Pliocene Boundary Event)
4.4.1.. End of Tropical Ecology - *End of Garden of Eden*
4.4.1..1. Destruction of Habitat/Ecology event
4.4.1..2. Tectonic motion event
4.4.1..3. Volcanism event

A major theme in the creation accounts of the bible is the recurring mass extinction followed by renewal of ecology.

Part 4 of Eon 4 is another mass extinction. This time, humans were present and were survivors.

Everyone has heard of the mass extinction of Noah that humanity has survived. But few have heard that the end of the Garden of Eden was an earlier mass extinction. It is clearly presented in the ancient scripture of the bible, but theologians have ignored it. It just did not fit with their limited understanding of the bible in the light of their traditional interpretation.

It is uncanny how the details of the bible and the discoveries of science line up. Yet, the traditional interpretation time scale versus Geologic time scale are so different that no one in their right mind would make the correlation.

On the other hand, ignoring the traditional interpretation, the bible has no time scale and insists that time is unimportant. Peter, in his creation account says time is much longer than we expect.

In the mass extinction at the end of the Garden of Eden, the prominent feature was volcanism and tectonic activity. In the bible that event is described as erupting volcanoes and lava flows. In the discoveries of modern science it is described using the phrase, "tectonic activity." (Tectonic activity includes erupting volcanoes and lava flows.)

This mass extinction ended the reign of mild tropical climate of this planet. The pole-to-pole tropics gave way to ice ages. Thorns and thistles dominated the surviving ecology—but that is to be discussed in the part 5 of Eon 4.

In the Everlasting-to-Everlasting Creation Account recorded by Moses in about 1446 BC, some thirty-five hundred years ago, Moses lists several repetitions of the mass extinction cycle. The first mass extinction in this list involved humans and their survival. This mass extinction happened thousands of years before the flood of Noah, mentioned later in the same account.

> [3a] *Thou causeth mortal humanity to return from destruction...*
> Moses, c. 1446 BC, *Everlasting-to-Everlasting,*
> *A Biblical Creation Account:* Psalms 90

In the Seven-Day Creation Account recorded by Moses in about 1445 BC, Moses makes reference to this mass extinction without mentioning it explicitly. Just after the verse giving the account of the creation of humanity, God is recorded as using the exact same language he used after the mass extinction of Noah's flood, thousands of years later—God told humanity to re-populate the earth. Some will argue that this language did not mean what it said; that Noah's flood was the only mass extinction in the bible. However, the language clearly is an exact copy of that used after the flood of Noah where it definitely meant to re-populate the earth after it had been populated before and was to be populated again. Therefore, it is a clear reference to an earlier population of humans before this specific mass extinction.

> [27] *So God created man in his [own] image, in the image of God created he him; male and female created he them.*
> [28] *And God blessed them, and God said unto them,* **Be fruitful, and multiply, and replenish the earth** *...*
> Moses, c. 1445 BC, *Seven-Day Creation Account,*
> *A Biblical Creation Account:* Genesis 1
> [1] *And God blessed Noah and his sons, and said unto them,* **Be fruitful, and multiply, and replenish the earth.**
> Moses, c. 1445 BC, Flood Account: *The account of Noah's Flood,*
> *A Biblical Creation Account:* Genesis 9

In the Ecology of Paradise Creation Account recorded by Moses in about 1445 BC, Moses gives more specific details of this mass extinction. A careful reading of the original Hebrew language reveals the driving from the Garden of Eden was not in space, but in time— Man continued to till the same ground, the ground from which he was taken. The destruction of the ecology of paradise was accompanied by volcanic (tectonic) activity

---

23   *Therefore the LORD God sent him forth (in time) from the Ecology of Paradise, to till the ground from whence he was taken.*

24   *So he drove out the man; and he replaced the Ecology of Paradise with erupting volcanoes, and hot lava flowing in every direction, to guard the way of the tree of life.*

Moses, c. 1445 BC, *The Ecology of Paradise,*
A Biblical Creation Account: Genesis 3

Even more clarity is given to this specific mass extinction and its position in the chronology of ancient history by King David in his *Chronological Order of Creation,* account written in about 1015 BC. Verse 32 below puts it in context.

Chronological Order of Creation

The Eon of Complex Life Forms. The Age of Monsters (Dinosaurs)
24*O LORD, how manifold are thy works!*
   in wisdom hast thou made them all:
   the earth is full of thy riches.
25*[So is] this great and wide sea,*
   wherein [are] things creeping innumerable,
   both small and great beasts.
26*There go the ships: [there is] that leviathan,*
   [whom] thou hast made to play therein.
27*These wait all upon thee;*
   that thou mayest give [them] their meat in due season.
28*[That] thou givest them they gather:*
   thou openest thine hand, they are filled with good.
Major Mass Extinction. Cretaceous/Cenozoic (Kt) Boundary Event
29*Thou hidest thy face,*
   they are troubled:
Thou takest away their breath,
   they die, and return to their dust.
Ecology Restored.
30a *Thou sendest forth thy spirit,*
30b *they are created:*
30c *And thou renewest the face of the earth.*

Era of Sustained Ecology.

³¹ᵃ The glory of the LORD shall endure for ever:
³¹ᵇ the LORD shall rejoice in his works.

Another Mass Extinction.   Miocene/Pliocene Boundary Event

³²He looketh on the earth,
   and it trembleth:
He toucheth the hills,
   and they smoke.

David, c. 1015 BC, *Chronological Order of Creation,*
A Biblical Creation Account: Psalms 104

An interesting statement is made by Isaiah in his creation account written between seven hundred BC and six hundred eighty BC. In his discussion of isostacy, he mentions the uplifting of the coastlands.

¹⁵ *Behold, the nations [are] as a drop of a bucket, and are counted as the small dust of the balance: behold, he lifts up the coastlands as though they were nearly nothing.*

Isaiah, c. 700-680 BC, *Early Planning and Detail,*
A Biblical Creation Account: Isaiah 40

This is interesting because there is a man made sea port thousands of feet above sea level in the Andes Mountains near Lake Titicaca. There is considerable variation among scholars as to the exact timing of the uplift, but most agree the area was at sea level near the end of the Miocene Epoch and was uplifted early in the Pliocene Epoch. Geologists have been reporting on discovery of widespread tectonic activity involving uplifting at or near the Miocene—Pliocene boundary.[59] Of a certainty, when members of the human race constructed the seaport, the area was at sea level. And now, it is thousands of feet above sea level. It is highly likely this uplift will eventually be associated with the tectonic activity, including uplift, associated with the destruction of the Garden of Eden.

---

[59] *Available References are abundant. Two examples are:*
Gregory-Wodzicki K. M. 2000 Uplift history of the Central and Northern Andes: a review. Bulletin of the Geological Society of America *112: 1091-1105*
Dickinson, J.A., Wallace, M.W., Holdgate, G.R., Gallagher, S.J. & Thomas, L., 2002, Origin and timing of the Miocene-Pliocene unconformity in southeast Australia, Journal of Sedimentary Research, *72, 317-332.*

4.5: Ecology Restored

4.5.   Ecology Restored, Global Cooling

4.5.1..   Era of restored ecology (Long time passes.)

4.5.1..1.   Climate turns colder and harsher - Global Cooling

4.5.1..2.   Thorns and Thistles ecology (carbon starved ecology)

4.5.1..3.   Expansion of humanity after habitat/ecology destruction -
                          Replenish

4.5.1..4.   Dedication of food chain

4.5.1..5.   Photosynthesis is the basis of the food chain for all forms of life.

4.5.1..6.   Pre-historic Past of Humanity

4.5.1..7.   Rate of Extinction exceeds Rate of Speciation
                          (Creation Complete)

4.5.1..8.   Long Time Passes. Pleistocene

4.5.1..9.   Rise of language and civilization

After[60] humanity survived its first mass extinction, life was a lot harsher. Thorns and thistles dominated the fields. The climate turned colder. They even had to wear clothes.

Scientists have come up with many theories as to why the climate has been colder and harsher since the end of the Miocene Epoch. Some speculate it is due to the changes in ocean current patterns due to the tectonic changes such as the uplifts.   Others speculate it is because there is less carbon dioxide in the atmosphere. Currently we are living in a time when the ecology is starved for carbon dioxide. To moderate the climate and restore the rainforests of yesteryear much more hydrocarbon must be recycled from the buried coalfields[61].

Scientists may have varied opinions as to what caused the world climate to change, but they accept the basic premise that it has. The fact that they realize it actually happened is only of recent origin. The fact that the Garden of Eden is a time rather than a place has not been discussed in scientific nor religious literature that TheOldScientist can find. As far as the politically correct scientist is concerned, this fact of a climate change from the paradise of the past is not confirming what the bible published thousands of years before.   As far as the

---

[60] *At this point in the chronology we have a period of time after an event and before another event in which the bible gives many descriptive details that have no chronological significance within that period. Notice that many detail numbers end in a digit after the double decimal indicating no chronological significance within that period of time.*

[61] *See,* Carbon Dioxide/Hydrocarbon Cycle. *Page 81, in the book, Eyewitness to the Origins*

politically correct scientist is concerned, it is an independent observation that disproves the bible. Little does he know it is only the erroneous traditional interpretation that is being disproved.

Again, it is uncanny how the details of the bible description and the discoveries of science line up. Yet the difference in time scales from the traditional interpretation versus the reality of what the bible says—and verified by modern science—are so different no one in their right mind would make the correlation. Unless, of course, if they were to ignore the traditional interpretation of the timing.

Yet, Moses plainly stated the facts over thirty-five hundred years ago—the climate and ecology would be harsher, and colder:

> *17* *And unto Adam he said,   cursed [is] the ground for thy sake; in sorrow shalt thou eat [of] it all the days of thy life;*
>
> *18* *Thorns also and thistles shall it bring forth to thee; and thou shalt eat the herb of the field; ...*
>
> *21* *Unto Adam also and to his wife did the LORD God make coats of skins, and clothed them.*
>
> Moses, c. 1445 BC, *The Ecology of Paradise,*
> *A Biblical Creation Account:* Genesis 3

Notice, in the new ecology and climate, they had to wear clothes. Earlier, in the old ecology and climate, they wore no clothes.

> *25* *And the man and his wife were both naked and were not ashamed.*
>
> Moses, c. 1445 BC, *The Ecology of Paradise,*
> *A Biblical Creation Account:* Genesis 2

Currently, the only place on this earth where people are not expected to wear clothes is in the tropics where there is no need for clothes to keep warm.

This change in ecology and climate ushered in a long era of stable ecology. During this long period of time, humanity flourished and civilization expanded. Notice, according to Moses in the *Seven-Day Creation Account,* that it was after this first human survival of a mass extinction that the formal dedication of the food chain took place. Previous to this, in the era before the destruction of the Garden of Eden, the climate was much more habitable and there was no need to "subdue it."

Re-population of earth after mass extinction.

1:28a *And God blessed them, and God said unto them, Be fruitful, and multiply, and replenish the earth, and subdue it: and have dominion over the fish of the sea, and over the fowl of the air, and over every living thing that moveth upon the earth.*

> The exact same language is used here as when ordering Noah and his sons to re-populate the earth.   That indicates that this is an event after a mass extinction as was in the case of the Expansion of humanity by Noah and his sons after the flood had destroyed their habitat/ecology.

The Formal Dedication of the Food Chain

1:29a *And God said, Behold, I have given you every herb bearing seed, which [is] upon the face of all the earth, and every tree, in the which [is] the fruit of a tree yielding seed; to you it shall be for meat.*

1:30a *And to every beast of the earth, and to every fowl of the air, and to every thing that creepeth upon the earth, wherein [there is] life, [I have given] every green herb for meat: and it was so.*

> Photosynthesis of water and carbon dioxide into hydro-carbon is the basis of the food chain for all forms of life.

Pre-historic Past of Humanity.

1:31a *And God saw every thing that he had made, and, behold, [it was] very good.*

> In stating **very** good indication is made of completeness of all phases of creation where each of the previous phases of creation were indicated to be complete by the same phrase without the **very**.

And the evening and the morning were the sixth day.

The creation is finished (complete).

2:1 *Thus the heavens and the earth were finished, and all the host of them.*

> This describes the part of the punctuated equilibrium cycle where rapid speciation after a mass extinction is replaced by sustained ecology. Currently, the rate of extinction exceeds the rate of supposed evolution by

123

many orders of magnitude. Science claims extinctions are at the rate of thousands a year. Science is hard pressed to show any examples of new species coming into existence.

The Present Era of Stable Ecology

²:² *And on the seventh day God ended his work which he had made; and he rested on the seventh day from all his work which he had made.*

We are living in an era of sustained, stable ecology. Note: Noah's flood was in the long period of rest, after the termination of creation as God had Noah preserve life forms rather than God re-create them after the flood. Contrary to most bible scholars' opinion, the flood of Noah was not such a big deal as previous mass extinctions. It did not require any species re-creation.

²:³ᵃ *And God blessed the seventh day, and sanctified it: because that in it he had rested from all his work which God created and made.*

Moses, c. 1445 BC, *Seven-Day Creation Account,*
A Biblical Creation Account: Genesis 1-2

In the chronology of the first creation account recorded by Moses in about 1445 BC, he emphasizes the long passage of time between the recovery from the first mass extinction survived by humanity and the flood of Noah:

Surviving a Mass Extinction:

³ᵃ *Thou causeth mortal humanity to return from destruction; and commandeth:*
³ᵇ *"Return, ye descendents of the human race."*

Long Passage of Time:

⁴ᵃ *For a thousand years in thy sight [are but] as yesterday when it is past, and [as] a watch in the night.*

The Flood of Noah:

⁵ᵃ *Thou carriest them away* ⁵ᵇ *with a flood; they go to sleep:*

Moses, c. 1446 BC, *Everlasting-to-Everlasting,*
A Biblical Creation Account: Psalms 90

Even the creation account by Peter, written in about 68 AD—patterned after that same Mosaic account—emphasizes the long passage of time during this era by repeating that same passage from that same first Mosaic creation account:

> 8a *But, beloved, be not ignorant of this one thing, that*
>
> 8b *one day [is] with the Lord as a thousand years, and a thousand years as one day.*

Peter, c. 64-68 AD, *Eternity-to-Eternity,*
A Biblical Creation Account: 2 Peter 3

It is curious that the *Chronological Order of Creation* account of King David only acknowledges all of human history from the end of the Garden of Eden to the coming future Judgment, including the era before and the era after the Flood of Noah by a simple casual three line mention in verses 32 through 34. After mentioning many scientific points of detail in exact chronological order—prehistoric events and detail—that were unknown to humanity until discovered by modern science, it is as though to mention known history was unnecessary, only worthy of a casual acknowledgment. Or maybe it is because the reign of humanity on this planet has been only during an insignificant portion of the total existence of this planet.

Mass Extinction (Garden of Eden).   Miocene/Pliocene Boundary Event
> 32a *He looketh on the earth,*
>> Destruction of Habitat/Ecology event/Miocene-Pliocene
>
> 32b *and it trembleth:*
>> Tectonic motion event/Miocene-Pliocene
>
> 32c *He toucheth the hills, and they smoke.*
>> Volcanism event/Miocene-Pliocene

Ecology Restored After Destruction of Garden of Eden.
> 33a *I will sing unto the LORD as long as I live:*
>> Era of restored ecology (Long time passes.) /Pliocene

Ecology Restored After Noah's Flood.
> 33b *I will sing praise to my God while I have my being.*
>> Long Time Passes. (Pleistocene)

Ecology Endures into Future.
> 34*My meditation of him shall be sweet: I will be glad in the LORD.*
>> Era of restored ecology Proliferation of civilization after
>> near extinction/Holocene

Another Mass Extinction   Holocene/Eternity Future Boundary Event

35a Let the sinners be consumed out of the earth,
> A Future Mass Extinction – The
> Great and Terrible Day of the Lord/Phanerozoic Future

35b and let the wicked be no more.
> Secret iniquities revealed (Judgment Day)/Eternity Future

Ecology Restored in Eternity Future.

35c Bless thou the LORD, O my soul.
> Human existence in eternity future/Eternity Future

35d Praise ye the LORD.
> God Existence In Eternity Future/Eternity Future
> David, c. 1015 BC, *Chronological Order of Creation,*
> *A Biblical Creation Account:* Psalms 104

There is no question that the bible published the fact that between the earlier mass extinction at the end of the Garden of Paradise, and the mass extinction at the flood of Noah, there was a great passage of time. This corresponds well with what modern science has been recently discovering concerning human history. There is no question that modern science had uncovered the fact that between the earlier mass extinction—at the end of the Miocene—that ended the ecology of paradise era, and the more recent mass extinction—at the end of the Pliocene epoch—of humanity and large mammals, there was a great passage of time.

4.6: Another Mass Extinction—Noah's Flood

4.6.   Another Mass Extinction
                          (Pleistocene/Holocene Boundary Event)
4.6.1..1. Mass Extinction of Large Mammals
4.6.1..2. Mass Extinction was in form of a flood (*Noah's*)

The Problem: No Mechanism—No Correlation.

In the late Pleistocene epoch, which ended about 11,000 years ago, there were drastic climate shifts. Also, within the past half century, modern science has become aware that there was, at the same time, a mass extinction of megafauna in the North American continent.

This should come as no surprise to a science student of the bible. That would correspond very well to the time of the mass extinction described in the bible in the form of Noah's flood.

In the case of the mass extinction of the end of the Garden of Eden it was the perceived time scale differences that prohibit anyone in their (politically correct) right mind from making the correlation to the Miocene/Pliocene boundary.

In this case there is no discrepancy in the timing of the event between the flood of Noah and the late Pleistocene extinction. This time it is the details that confound the correlation. The bible describes a great flood as being responsible for the mass extinction. Modern science does not recognize any mechanism that would account for enough water to inundate the continents, and no mechanism that would explain where the water went after the inundation.

The Bias Against Considering any Possible Mechanism:

In a way, this lack of a mechanism to produce vast amounts of floodwater is a relief for modern science. They can immediately dismiss as pseudo-science any attempt at correlation between their facts and what they consider to be the myths of the bible.

Many religious attempts have been made to explain Noah's flood. Many pseudo-science explanations and hypotheses have been proposed. None have held water—pun intended. So ridiculous are the myriads of proposals that it is embarrassing to the intelligence of the average student of science, let alone any professional scientist. As a result, modern science refuses to consider any mechanism that would produce enough water to overflow the continents. Nor do they consider any mechanism that would dispose of the excess water after the inundation.   It is obvious to them that any such proposed mechanism is just a ploy to make the bible seem to be scientific.

However, this produces a bias against any research that has to do with vast inundation.

An Example of the Bias: J. Harlen Bretz.

In the early part of the twentieth century, a geologist by the name of J. Harlen Bretz devoted his life to study of the scablands of the Northwest of the United States.   The topography was obviously shaped by a vast flood sometime in the prehistoric past.   When he published the obvious, he was openly and vigorously ridiculed and discredited.   As Sean D. Pitman[62] put it,

> "Sometime before 1927 geologists were catching on to the seriousness of what Bretz was suggesting. If true, Bretz's theory would undermine the very foundation of Uniformitarianism.   Just as anticipated, the general outcry against any hint of a catastrophic model was very loud indeed. In fact, there was a very strong desire to publicly discredit and humiliate Bretz."

For many years the debate raged. In 1940, things began to slowly change. It was discovered that there was in fact a great prehistoric lake Missoula that was held back by a glacial ice dam. That ice dam had failed catastrophically and provided the floodwater needed for Bretz observations. Today it is universally accepted that floodwater did in fact cause the topography described by Bretz, but only because of the acceptance of a mechanism to provide that water. Finally, in 1979, when Bretz was in his late 90's he was honored for his work.

Of Course, the flood of Bretz was not the flood of Noah. The flood of Bretz only lasted about 48 hours as Lake Missoula catastrophically emptied.   The water rose fast, was moving rapidly, and eroded massively. The flood of Noah rose slowly, lasted longer, and receded quietly, leaving much less concentrated evidence of a catastrophic event.

---

[62] Sean D. Pitman, M.D., 2004,   J Harlen Bretz And The Great Scabland Debate, © April 2004

http://www.detectingdesign.com/harlenbretz.html   retrieved from the internet April 19, 2007

Even so, the fear of even considering anything that might possibly be interpreted as confirming the bible is demonstrated in the attitudes and actions of his colleagues. Of course, this is no worse than what the theologians did to Galileo. It is just a fact of life. As J. Harlen Bretz himself said in 1928, *"Ideas without precedent are generally looked upon with disfavor and men are shocked if their conceptions of an orderly world are challenged."*

In the case of the flood of Bretz, the hidden agenda was evident in the resistance to pursue the truth. The truth was acknowledged only when the mechanism was discovered.   There was in fact, a catastrophic flood. But to the relief of the scientific community, it was not the flood of Noah.

The greatest trump card held by the resistors to an inundation on a continental scale, is the lack of a viable mechanism to explain where the great quantity of water came from and where it went. The Bretz experience is only one of many that demonstrate the reality of resistance to any research on any proposal that might lead to a mechanism for a large scale flood—or to any other departure from politically correct science.   Any such departure will be avoided, resisted, vigorously ridiculed, and discredited.

The Result of the Bias:

Since the correlation of the time of the latest, mass extinction, and the flood of Noah is so obviously matching, no scientist in his right mind would dare publish any hypothesis that inundation were in any way responsible for that mass extinction—or for any other mass extinction of the past. Any detail that would indicate in any way agreement between the flood of Noah and the scientific version of the mass extinction would have to pass extreme scrutiny or have the scientist doing the publishing risk ridicule and career.

Theologians too, have a problem in making the correlations. To be sure, much of what they have considered to be allegory or theobabble in the bible will have to be taken as literal. Much of what they have taken literally will have to be converted to allegory. That will be a difficult transition—but that subject is too involved to discuss here.

Hope for the Future:

Even though, at this time there is no agreement among scientists as to the mechanism that caused this latest mass extinction, there is hope. For many years many scientists have been laboring under the hypothesis that it was humans over hunting that caused the North American Large Animal Extinction. More recently that hypothesis has come under serious scrutiny. A recent article[63] states:

> "Scientists have been picking over the bones and evidence for more than three decades but cannot agree on what caused the extinction of many of the continent's large mammals. Now, in two new papers, a University of Washington archaeologist disputes the so-called overkill hypothesis that pins the crime on the New World's first humans, calling it a "faith-based credo" that bows to Green politics."

> "While the initial presentation of the overkill hypothesis was good and productive science, it has now become something more akin to a faith-based policy statement than to a scientific statement about the past," said Donald Grayson, a UW anthropology professor."

While the debate as to the cause of the extinction continues[64], at least modern scientists agree that there was such a mass extinction about the same time as what they consider to be Noah's mythical flood. It is interesting and encouraging to observe that scientists themselves acknowledge their own faith based tendencies.

A Possible Mechanism: The Lithologic Cycle

A possible mechanism for the episodic nature of the cycle of mass extinctions was proposed in 1996 by The Old Scientist[65]. Without fanfare and without widespread distribution it was blocked from mainstream science.[66] You see, any scientific research that might

---

[63] *Grayson, D. K., Oct. 24, 2001,* Blame North America Megafauna Extinction On Climate Change, Not Human Ancestors, *Article edited by Joel Schwarz University of Washington* http://uwnews.washington.edu/ni/article.asp?articleID=2654
Source: unews.org/University of Washington

[64] *Alroy, J. 2001. A multispecies overkill simulation of the End-Pleistocene megafaunal mass extinction. Science 2001 292: 1893–1896.*
Grayson, D. K., and D. J. Meltzer. 2002. Clovis hunting and large mammal extinction: A critical review of the evidence. Journal of World Prehistory *16: 313–359. Grayson, D. K., and D. J. Meltzer. 2003. A requiem for North American overkill.* Journal of Archaeological Science *30: 585–593.*

[65] *Frederick, M. B., 1996,* Origin of the Continents: An Introduction to the Theory of The Lithologic Cycle, *Max B. Frederick Publishing, 146 Laurel St., Central Point, Oregon 97502*

possibly result in any verification of the truth of the bible could not possibly be published in any peer reviewed scientific journal. No scientist who values his status, position or employment in the scientific community would dare even publish such a peer review—even if the science behind the research were sound.

Examples of such censorship are found among scientists who have dared to attempt a peer review of any article on "intelligent design." The opponents of such research even glory in the fact that no article containing pro intelligent design evidence has ever been published in any peer reviewed journal. In fact, proponents of spontaneous generation went so far as to take a case to the United States federal court to get a ruling that a public school district requirement for science classes to teach intelligent design as an alternative to evolution was a violation of the Establishment Clause of the First Amendment to the U.S. Constitution.[67] A pro spontaneous-generation article in the encyclopedia Wikipedia exhibits pride in the fact that," During the trial, intelligent design advocate Michael Behe testified under oath that no scientific evidence in support of the intelligent design hypothesis has been published in peer-reviewed scientific journals."[68] You can interpret that statement as either the fact that no such evidence exists, or that there is censorship. Obviously, pro spontaneous-generation advocates insist on interpreting it to mean that no such evidence exists. However, denial of the truth does not make it false. Truth cannot be established nor eradicated by consensus. Science by court ruling returns science to the dark ages of pseudo-science.

Such is the world we live in.

Therefore, with this warning against mainstream refusal to even consider such a theory, a brief summary of the Theory of The Lithologic Cycle is re-published here anyway. For a complete version obtain a copy of the original publication from the Internet.[69]

---

[66] *The paper was submitted, but no acknowledgment of that submission was ever received.*

[67] *Ruling, Kitzmiller v. Dover Area School District, Case No. 04cv2688. December 20, 2005*

[68] *Intelligent design. (2007, April 18). In* Wikipedia, The Free Encyclopedia. *Retrieved 16:02, April 18, 2007, from*
    http://en.wikipedia.org/w/index.php?title=Intelligent_design&oldid=123833774

[69] *www.EyeWitnessToTheOrigins.com follow links to, The Lithologic Cycle Book.*

If it were not for the fact that the Theory of The Lithologic Cycle has the side effect of providing a mechanism for explaining where the water of Noah's flood came from and where it went after the flood, there would be serious consideration by the mainstream scientific community.

The Theory of The Lithologic Cycle —A brief Summary

According to the Theory of The Lithologic Cycle, tectonic activity—the relative movement of the continents—occurs at two very different speeds—creep speed and earthquake speed. Not only is movement uniform over eons of time—at creep speed—as in the uniformitarianism sense, but movement also occurs in episodic bursts at earthquake speed. In recent years it has been discovered that continental and oceanic plates of the earth's crust move relative to each other. The observed continuous movement is generally very slow—on the order of a few centimeters per year at most. Occasionally, tectonic plates will be seen to move at a very rapid speed—about the speed of a fast jog—during earthquake activity. But that is a rare event. This rapid movement is not very far and does not last long. For the most part, the uniformitarianism philosophy of continental relative movement is observed to be the rule.

However, in the geologic record, there are observed evidences of mountain building episodes. No satisfactory mechanism has yet been proposed to explain why the record shows long periods of quiescent slow movement of the continental plates, interrupted by rare episodes of mountain building—until the Theory of The Lithologic Cycle was proposed in 1996.

The Theory of The Lithologic Cycle outlines a series of phases of a continuously repeating cycle.

There are the four phases of the Lithologic Cycle:

Phase 1: **Isostatic Balance Phase**. Long in duration—many thousands of years. Uniformitarianism is observed. Civilization arises and eyewitness accounts of the other phases become myth.

Phase 2: **Trigger Phase**. The start of a relatively rapid succession of events.

Phase 3: **Rapid Turnover Phase**. Rapid relative movement of continental plates—duration on the order of a month—sinking of more dense ocean floor tectonic plate into the interior of the earth—deposition of sialic material under continent—sea level temporarily rises, overflowing onto continent as the continents are temporarily drug down by the ocean floor sinking beneath the continents.

Phase 4: **Isostatic Rebound Phase**. The long missing explanation for the observed periodic mountain building episodes—duration on the order of a year—the return to isostatic balance as new material deposited at the roots of mountain range buoy up the mountains above—to levels higher than immediately previous to phases 2 and 3—the receding of the ocean waters back into the ocean basins.

How could this be? The earth is so stiff and rigid that such a catastrophic event would be impossible.

Quite the contrary. If you could construct a scale model of the planet earth, it would be so fragile you could not touch it without catastrophic results.

Visualize This:

Imagine a scale model of the earth a half-inch smaller than a regulation ten-pin bowling ball. That would be a scale of one-inch equals a thousand miles. The earth is about eight thousand miles in diameter. The scale model would be about eight inches in diameter. Except for the solid core and the crust, the earth is mostly a thick liquid. The model would weigh almost twenty pounds—more than any bowling ball. The scale model—if you also scaled the strength of materials—would be extremely fragile. Its scale viscosity would be more liquid than water.   The crust would be only about five thousandths of an inch thick and less rigid than pond scum. The crust would be so fragile you could not set the scale model earth on a table. If you tried to set it on a table, it would immediately collapse and flatten out like a puddle of water. You would have to suspend it in a weightless environment like outer space. You could not even touch it without destroying the pond scum crust. Even worse, any plate of the crust under the oceans, being denser than the liquid rock under that crust plate, would tend to crack, break, and sink into the interior. Realize that the earth is so flexible that the effect of the gravity of the moon causes a twice-daily tidal movement in the actual hard rocks. The solid surface of the earth rises and falls about a foot. It happens twice a day. That would be equivalent to the gravitational attraction of a billiard ball twenty feet away from the scale model earth, causing a deformation of the scale model earth, just by being there. You could not even come into the presence of the scale model earth without destroying it by the gravitational attraction of just your body. The deformation of the scale model earth would crack the pond scum, causing it to sink into the interior, destroying civilization.

Hard to imagine something that fragile? Try to imagine the strength of rock in a vertical cliff. Suppose the cliff were a mile high and had an overhang of maybe a few hundred feet. You would not dare walk under it as it would collapse under its own weight. Now try to imagine such a cliff with an overhang of as much as a mile. Impossible. Rock is not that strong. Now scale that down to the size of the bowling ball and that one mile overhand would be a cliff only a thousandth of an inch high with an overhang of only one thousandth of an inch—with such weakness that it would fall to imperceptible smoothness. Suddenly you can imagine a scale model of the earth so smooth you could not even see any roughness. All this fragileness would be due to the actual lack of strength of materials. Such is the planet earth

Compare that to a drop of water sitting on the kitchen counter. It sticks up like a bubble due to the strength of the surface tension. That surface tension is unimaginably stronger than the strength of the crust of the earth.

In fact, the strength of the crust of the earth is so weak that a six-mile diameter meteorite striking one side of the earth actually caused deformation on the opposite side of the earth. And it caused a mass extinction of enormous proportions causing a complete change of the ecology of the earth. Now, on the size of our scale model earth, the size of that meteorite would be only a little larger than the diameter of a human hair. And the speed with which it hit the model earth was only about one fiftieth the speed of a bullet. The earth had to be very fragile.

In life size, the estimated speed of the meteorite that struck the earth, when it impacted the earth at the end of the age of the dinosaur, was about 11 km/sec. That is pretty fast and is unimaginable in the damage caused, but in the larger perspective, if it had missed the earth going at that speed—over thirty times the speed of sound—it would have taken over fifteen minutes to go past the distance of the diameter of the earth.

Now, back to the Lithologic Cycle Theory.

Phase 1: Isostatic Balance Phase

We are currently living in the Isostatic Balance Phase. Isostatic balance is where the weight of a vertical column of mass of one horizontal area is the same as any other equal size horizontal area. The mountains are less dense and therefore rise higher in the mantle. The ocean floor is denser and therefore sinks lower into mantle. The Isostatic Balance Phase is a relatively long period of time compared to the other phases.

However, during this phase there is a build up of stress in preparation for the rapid succession of the other phases. The earth is made up of concentric spheres of mater. The more central sphere is the densest and the outermost—the atmosphere—is the less dense. Each concentric sphere is less dense than the one below. There is one exception. Therein lies the mechanism that drives the cycle. Where a continent is made up of rocks that are less dense than the more liquid rock material below and therefore floating so that it cannot sink, the ocean floor is another story. The rock that makes up the ocean floor is the solidified versions of the more liquid (partially melted) rock matter immediately below. When rocks cool and solidify, they shrink. This makes them denser than melted the rock material below.

Phase 2: The Trigger Phase

When enough stress builds up, or when some outer influence disrupts the delicate crust of the earth, such as a meteorite impact, a sinking is triggered. This puts us into phase two, the trigger phase. When broken up, that denser solidified rock of the ocean floor tends to sink into the liquid rock below it. This is unlike less dense solidified water—ice—floating on liquid water. The only thing preventing this sinking of the ocean floor is the delicate bridging over of the solidified rock. If it breaks on a large scale, it will sink. And it sinks quite rapidly compared to the creep speed we observe today.

Phase 3: The Rapid Turnover Phase

This puts us into the rapid turnover phase. The Rapid Turnover Phase relieves the strain that is built up during the long duration of the Isostatic Balance Phase. The Rapid Turnover Phase is an event where the old ocean floor rapidly slides sideways, and down, sinking into the mantle below a continent in an avalanche type event. This generally occurs at the edge of a continent with the old ocean floor moving toward the continent in the same direction as the slow subduction seen during the Isostatic Balance Phase. Once started, the avalanche continues until the strain is relieved and/or inertia drives the plates no further. The speed again returns to a slow creep. The maximum duration of this phase when the plates are moving at earthquake speed comparable to a fast jog, is on the order of magnitude of a month. Any slower movement of the avalanche would allow the sliding surfaces to again lock down to creep speed.

To better understand the difference between the two tectonic plate movement speeds—creep speed and earthquake speed—consider the difference between static versus sliding friction. Basic to friction is the phenomena of a difference in resistance to static versus sliding friction. When two objects are rubbing against each other, when the motion is stopped (static) it is harder to get it moving than it is to keep it moving once it is started. This can be easily illustrated by rubbing your finger and thumb together when they are dry and not lubricated. Slowly rub them together while pressing them together. Press them together harder and harder until the smooth sliding motion is replaced with a jerking, start/stop type sliding. The start/stop jerky motion is because once they start sliding, they keep sliding until they have caught up to where the stress is relieved and they again appear to stick. What appears to be the stick time is the slow creep speed of a few centimeters per year, and the rapid motion is the more rapid earthquake speed, about the speed of a fast jog.

During this sinking of the old ocean floor, new molten rock material from below rises to the surface below the ocean along the trailing edge of the sinking tectonic plate. As the old plate moves away, it leaves a trail of newly solidified ocean floor to replace the sinking old ocean floor. This new ocean floor is less thick than the old and has less average density. Therefore, to be in isostatic balance, it rises higher, causing the ocean basin to be less deep, causing the ocean water to overflow the continental areas. This unloading of the weight of water from over the ocean floor causes further rise of the ocean floor and the weight of the water loading on the continent causes more sinking of the continental area that is inundated, thus further inundating the continent and compounding the problem.

Phase 4: Isostatic Rebound Phase

Fortunately, the water loading on the continental areas is a temporary problem. In phase 4, as the new ocean floor cools, and thickens, due to the cool water above, the ocean basin again deepens and the water returns from the continents. Thus, everything is again stable.

Meanwhile, the avalanche has stopped and the less dense, lighter continental rocks that had been eroded to cover the ocean floor—and were now carried down under the continents as though the avalanche were a conveyor belt—float upwards and are added to the roots beneath the mountains along the edge of the continent. These lighter rocks beneath the edge of the continent are buoyed up to provide lift needed to explain mountain building episodes of the past. Thus, there grows a new range of mountains along the coastal area. Sometimes we see the coastlines are uplifted as along the Pacific coast of South America where we find a man made seaport uplifted nearly 12000 feet above sea level. Sometimes, volcanoes and lava flows release this newly melted continental material to the surface. Thus, the majority of volcanic activity is observed not far inland around such areas as the pacific coast.

Conclusion: Science has a lot to yet discover

Whether or not the Lithologic Cycle Theory is the final answer, modern science has a lot to discover before there is a satisfactory mechanism to explain the repeating pattern of mass extinctions as seen in the record of both the geologic column and in the ancient scriptures of the bible.

4.7: Ecology Restored
4.7.   Ecology Restored/Epoch of Secular Human History

Many young Earth Creation Science advocates insist that evidence is everywhere that the flood of Noah caused great devastation. They insist that evidence is everywhere in all the surface features of the earth. In reality, the flood of Noah was a benign flood that rose relatively quietly, covered vast areas of land, for a short while, and then quietly receded. Such an event would not require much time for ecology to return and heal the scars. There are a few examples of rushing water from great floods such as the scablands of the Columbia River basin causing damage that has persisted for centuries. On the other hand, scars from what we would consider catastrophic events such as the eruption of Mt. Saint Helens, have healed in such a short time that scientists were amazed.

The fact is, there is little, conspicuous, undeniable evidence of a widespread, continental inundating, flood near the end of the Pleistocene epoch. The account in the bible says Noah and his sons went to farming immediately after the flood receded. Farmers of the central valley of California return to farming the next year after a devastating flood. Little time is required for ecology to be restored after a widespread, inundating, slowly receding flood. To be sure, there may be evidence left after a widespread inundation, but it looks like any other inundating flood, of which there are many. In reality, a scientist would expect to see little if any evidence of the flood of Noah that would stand out as undeniable evidence of a continental inundation. It may have been a big deal to Noah and his family. Very little written record of human civilization before that event survives. It may have wiped out a great amount of population, civilization, and wildlife. It may have uplifted a seaport in South America to an elevation of over two miles. But in the overall view of all history, it was one of the minor events. Archaeology is beginning to find remnants of civilization dating to before that event.

Even the bible describes it as not one of the more significant extinctions. In contrast to earlier mass extinctions that required new species to be created to repopulate ecology, enough species survived Noah's flood so that a new wave of species creation was not necessary.

138

Since that last minor mass extinction, civilization has flourished and spread around the globe—and eyewitness reports of that event have come to be dismissed as religious myth.

4.8: Era of Restored Ecology

4.8.   Era of restored ecology
                              Proliferation of civilization after near extinction
4.8.1..   Beginning times/Current Era Boundary
4.8.2..   Living in Present
4.8.2..1.   "End Times" Philosophy (false)
4.8.2..2.   Uniformitarianism Philosophy (false)
4.8.2..3.   Logic of God, Laws of Physics (true)
4.8.2..4.   Present time/Future time boundary
4.8.3..   Living In Future
4.8.3..1.   Time of Sustained Ecology.

Archaeology has a dilemma: Most archaeologists hold to the belief that humans have existed on this earth for over twenty times the amount of time it took for a written record of civilizations to develop. Others extend that time by a factor of ten or even a hundred. As a consequence, they imagine many advanced civilizations have come and gone—without a trace of writing—in the first ninety-five percent of the existence of humans. On the other hand, they see written records of civilization beginning only a few thousand years ago. In the last five percent of the time that humans inhabited the planet earth, writing appeared, fully developed and in several different languages. There is something amiss. Either recording in the form of writing did not exist, or evidence of it was destroyed. It is highly unlikely that after 95 percent of the time of the existence of humans, that simultaneously in different parts of the world, in different languages, and different cultures, writing would suddenly simultaneously be independently developed. But that is exactly what the scientific evidence shows.

In about ten thousand years the population of humans exploded from just a few—the brink of extinction—to enough to overwhelm the ecology of the whole planet. In the at least two hundred thousand years before that, it should have happened twenty times over. What are we missing? Did it fail to happen, or did it happen with most of the evidence destroyed by inundation?

In either case, the written record from secular human history and the record from the bible after the account of the flood are consistent and in agreement. There was an explosion in both population and technology, as described in both sources.

In addition, the bible predicted how the philosophy of science would go during this era.

Philosophy of Science:

Peter, about 68 AD, wrote in his creation account, concerning the coming uniformitarianism philosophy of science in this era of restored ecology.

Uniformitarianism:

During this era, there would arise a dominant philosophy that would be used to develop a new theory of spontaneous generation when their old one failed.

> 2Pe 3:3,4   *Knowing this first, that there shall come in the last days scoffers, walking after their own lusts, And saying, Where is the promise of his coming? for since the fathers fell asleep, all things continue as [they were] from the beginning of the creation.*
>
> Peter, c. 64-68 AD, *Eternity-to-Eternity,*
> *A Biblical Creation Account:* 2 Peter 3:1-18

This passage is a statement of the Principle of Uniformitarianism, the accurate prediction that it would arise, and how it would be used.

The Principle of Uniformitarianism is a statement of the belief that there were no catastrophic events in the history of development of the earth. All things that we see operating today to shape the landscape are the same forces, acting in the same way in the past, over extended periods of time that result in the landscape as we see it today.

Paraphrased, this verse could be restated: "A time in the late 1700's shall come when James Hutton, a Scottish physician and gentleman farmer, the founder of modern geology shall publish his, "*Theory of the Earth*" promoting the Principle of Uniformitarianism. And scoffers [evolutionists] shall pick it up as a mantra against Christianity, saying, we don't need god, all we need is time, great lengths of time."

Old Earth vs. Young Earth Debate a Stumbling Block.

Peter goes on to say, in the next verse that it is not the age of the earth that is the issue, it is the cause that is important. Peter's implied logic between verses 4 and 5 is that even though the old earth opinion is used as a mantra against Christianity, there is some validity to the old earth concept and to deny it would be to deny reality. Later Peter argues that if we are not clear about such hard to understand things, we too may be lead away to our own destruction. (v. 16,17)

> 2Pe 3:5 *For this they willingly are ignorant*
>> Willing ignorance is confirmatory bias. This is a reference to the writings of Paul discussed in the next section. They fully understand the complexities of creation of the heavens and earth, the evidence for design, but refuse to give the creator God credit because that is the way they are predisposed to believe.

> *of, that by the word of God*
>> In the original language, this is the intelligence that created the universe, the laws of God, i.e. laws of physics.

> *the heavens were of old,…*
>> Yes, the heavens are old.

2Pe 3:15   *... Paul also according to the wisdom given unto him hath written[70] unto you;*

2Pe 3:16 *As also in all [his] epistles, speaking in them of these things; in which are some things hard to be understood, which they that are unlearned and unstable wrest, as [they do] also the other scriptures, unto their own destruction.*
> This is a quote from Psalm 93. Psalm 93 was the basis of the inquisition misinterpretation of scripture that led to the condemnation of the truth of science that Galileo was forced to recant to the embarrassment of the Catholic Church for three hundred sixty years. Also, consider Genesis 1:3-2:3, the misinterpretation of which is the basis of the embarrassing pseudo science popular in religion of today.

---

[70] *Paul 2: c. 57 AD. See:* Paul's discourse on False Theory, *Unprovable Mindset—False Theory*, on next page

2Pe 3:17   *Ye therefore, beloved, seeing ye know [these things] before, beware lest ye also, being led away with the error of the wicked, fall from your own stedfastness.*

> Every year hundreds, if not thousands of our best and brightest young people fall from the steadfastness of their faith when they go off to college and find that their faith in their parents religion, based on the pseudo science of the Young Earth Creation Science Movement is no more true than Santa Clause or the Easter Bunny.
>
> Peter, c. 64-68 AD, *Eternity-to-Eternity,*
> *A Biblical Creation Account:* 2Peter 3:1-18

Unprovable Mindset—False Theory

The writing of the bible came to an end nearly two thousand years ago. Since that time, the philosophy of science has followed the pattern set forth by Paul in his creation account, *Discourse on False Theory.* Modern science has become bent on proving the unprovable theory that spontaneous generation is responsible for all life—that there is no god—that we are responsible to no one—That we are just a product of accident. As a result, our value of life and treatment of one another has suffered. The attitude has developed that if it feels good, do it because all we can look forward to is death:

18a Because:

The wrath of God from heaven is upon every human
    who holds irreverence and injustice of the truth.

19a Because:

That which is known of God is apparent to them.

Because:

God hath made it apparent unto them.

20a By the things God did that were not seen,
    from the creation of the cosmos,
    to the achievements now being revealed,

His supposedly imperceptible eternal power and divine nature is clearly seen.

They become defenseless.

21a Because:

Having the knowledge of the God, they did not esteem him as God nor be thankful.

Instead:

They became vain in their logic and the intelligence of their heart is dim.

> [Paraphrase: The dimwits!, they refuse to figure it out from the evidence.]

22a Alleging themselves to be wise, they are made stupid,

23a And they exchanged the glory of the immortal God
> *for images of mortal humans, flyers, quadrupeds, and reptiles.*

24a **Therefore:**

God gives up on them,
> *(on their inner feelings,*
>> *into dirtiness of their bodies being devalued in them)*
>> *25awho exchange the truth of God, and*
>> *revere the false theory, and*
>> *offer divine service to the creation*
>>> *rather than the one creating,*
>>> *who is blessed into the eons.*

AMEN

PS:

> *26a As a result, God gave them up unto passions of negative value: for even their females exchange the natural use into the unnatural:*
>
> *27a And likewise also the men, leaving the natural use of the female, were burned out in their craving of one another; males in males, the indecency acting, and receiving in themselves the just wages for straying.*

PPS:

28a And in proportion to their not testing their knowledge of God,
> *God gives them over into the untestable theory,*
> *to be trying to do that which cannot be done.*
> [trying to prove the unprovable.]

PPPS:

29a Having been filled with all unrighteousness, fornication, wickedness, covetousness, maliciousness; full of envy, murder, debate, deceit, malignity; whisperers, 30a Backbiters, haters of God, despiteful, proud, boasters, inventors of evil things, disobedient to parents, 31a Without understanding, covenant breakers, without natural affection, implacable, unmerciful: 32a Who knowing the judgment of God, that they which commit such things are worthy of death, not only do the same, but have pleasure in them that do them.

Paul c. 57 AD, *Discourse on False Theory*
A Biblical Creation Account: Romans 1:15-32

The recorded history of this era in which we now live bears out the validity of these predictive writings written nearly two thousand years ago.

Turn on the television randomly to almost any random channel at almost any time and what you see is accurately described somewhere in verses 29 – 32 above, written about two thousand years ago as a prediction of the nature of entertainment today.

4.9: Future Major Mass Extinction—Day of the Lord

4.9.   End of Era of Complex Life Forms Event
4.9.1..1 There will be an end.
4.9.1..2 A Future Mass Extinction – Great and Terrible Day of the Lord.

Most theologians are skeptical when first confronted with the concept that the bible describes multiple mass extinctions, and with the concept that the bible describes it as a pattern that will continue into the future. They are biased by the traditional interpretation of the bible based on antique Greek science that did not recognize a pattern of past mass extinctions.

However, they are familiar with a coming great and terrible day of the Lord. When it is pointed out, it is easy to see that this is another, future, mass extinction. With careful study, one comes to the realization that this is just a continuation of a well-established pattern presented in many of the creation accounts of the bible. Once seen, the pattern is hard to deny.

In the first creation account of Moses, recorded in 1446 BC, he describes the final mass extinction of humanity just before his description of the coming judgment. In speaking of the "descendents of the human race," he says:

> ⁶ᶜ *in the evening it is cut down, and withereth.*
> ⁷ *For we are consumed by thine anger, and by thy wrath are we troubled.*
>
> Moses, c. 1446 BC, *Everlasting-to-Everlasting,*
> *A Biblical Creation Account:* Psalms 90

Peter's Eternity-to-Eternity creation account recorded in about 68 AD, is based on that ancient Everlasting-to-Everlasting account redorded by Moses. It follows the pattern in the Moses account. In this parallel account, Peter says it this way:

> ¹⁰ *But the day of the Lord will come as a thief in the night; in the which the heavens shall pass away with a great noise, and the elements shall melt with fervent heat,*
>
> Peter, c. 64-68 AD, *Eternity-to-Eternity,*
> *A Biblical Creation Account:* 2Peter 3

King David about 1015 BC foretold of the coming continuation of the pattern of mass extinction and destruction, then renewal of life.

> ²⁵ᵃ *In ages past*

25b *thou hast laid the foundation of the continents:*

25c *and the heavens [are] the work of thy hands.*

26a *They shall perish,*

26b *but thou shalt endure:*

26c *Yea, all of them shall wax old like a garment;*

26d *as a vesture shalt thou change them,*

26e *and they shall be changed:*

27a *But thou [art] the same,*

27b *and thy years shall have no end.*

28a *The children of thy servants shall continue,*

28b *and their seed shall be established before thee.*

David, c. 1015 BC, *Ages Past and Future,*
A Biblical Creation Account: Psalms 102:25-28

John, about 90 AD, in his creation account in the book of Revelation says it this way:

10:6 And sware by him that liveth for ever and ever, who created heaven, and the things that therein are, and the earth, and the things that therein are, and the sea, and the things which are therein, that there should be time no longer:

10:7b the mystery of God should be finished,

John, c. 90-96 AD, *Time and Eternity,*
A Biblical Creation Account: Revelation

Joel adds great detail to the description of the coming Day of the LORD, and actually ties it in to the biblical pattern of mass extinctions by mentioning the similarity to the destruction of the Garden of Eden.

"... the day of the LORD cometh, for [it is] nigh at hand; A day of darkness and of gloominess, a day of clouds and of thick darkness, as the morning spread upon the mountains: a great people and a strong; there hath not been ever the like, neither shall be any more after it, [even] to the years of many generations. A fire devoureth before them; and behind them a flame burneth the land as the Garden of Eden before them, and behind them a desolate wilderness; yea, and nothing shall escape them... The earth shall quake before them; the heavens shall tremble: the sun and the

moon shall be dark, and the stars shall withdraw their shining:"

Joel, c. 835-796 BC, *The Day of the LORD,*
A Biblical Creation Account: Joel 2:1-3,10

Joel compares the end of the Garden of Eden of the past with the terrible Day of the Lord of the future. He emphasizes the separation in time between the two as being almost forever, to the years of many generations. When comparing the end of the Garden of Eden with this, and other[71] biblical descriptions of the "Day of the Lord," it becomes obvious that they are both similar events accompanied by violent tectonic (volcanic) activity, and are separated in time by many thousands of years. One is in the ancient pre-historic past, the other at some unknowable future time that will come upon us as a thief in the night. The one is recorded by God in the rocks[72] of the geologic column; the other is predicted by the repeating nature of such events recorded in the same rocks, so as to leave even the modern day scientist without excuse.

---

[71] *See commentary on Peter's* Eternity-to-Eternity *creation account page 427 Eyewitness to the Origins,* Available at, *www.EyewitnessToTheOrigins.com*

[72] *End of tropical ecology at Miocene/Pliocene Boundary Event, the North American large animal extinction event with tectonic activity.*

Eon 5: Eon of Eternity Future

5.0. Eon of Eternity Future

5.0.0.1. Judgment
5.0.0.2. New Heaven & Earth.
5.0.0.3. Eternity Future Will Exist.
5.0.0.4. Human Existence In Eternity Future.
5.0.0.5. God Existence In Eternity Future.

Modern science has little to say concerning the eon of eternity future. It is outside the realm of science.

The realm of the bible, however, is the realm of eyewitness. And that eyewitness has demonstrated credibility by publishing knowledge of scientific facts and events long before the discovery of those same scientific facts—long before those scientific facts and events were known to humanity. That eyewitness testifies to items outside the realm of science, such as future life after death. Therefore, credibility of testimony concerning the eon of eternity future is exclusively the realm of the eyewitness testimony of the bible.

Even so, that testimony is entirely within the reality of what science could reasonably be expected to predict. Specifically, the continued repetition of the pattern of punctuated equilibrium. The bible speaks of future judgment with destruction, followed by another, future era of renewed existence.

King David's 1015 BC *Chronological Order of Creation* Account ends with these final four points of detail:

> **The future mass extinction ending Eon 4:**
> *35a* *Let the sinners be consumed out of the earth,*
> **The Coming Judgment and destruction**
> *35b* *and let the wicked be no more.*
> **Human Existence in Eternity Future (New Heaven and New Earth)**
> *35c* *Bless thou the LORD, O my soul.*
> **God Existence In Eternity Future**
> *35d* *Praise ye the LORD.*
> David, c. 1015 BC, *Chronological Order of Creation,*
> *A Biblical Creation Account:* Psalms 104

Since this account is a chronological ordered list of events from eternity past on beyond the present, these points can only be interpreted as future.

These same points are found in the 1446 BC, *Everlasting-to-Everlasting* account by Moses:

The future mass extinction ending Eon 4:
[7] *For we are consumed by thine anger, and by thy wrath are we troubled.*

The Coming Judgment and destruction
[8] *Thou hast set our iniquities before thee, our secret [sins] in the light of thy countenance.*

Human Existence in Eternity Future (New Heaven and New Earth)
[9a] *For all our days are passed...* [10e], *and we fly away."*

God Existence In Eternity Future
[2d] *even from Everlasting-to-Everlasting,*
[2e] *thou [art] God.*

> Moses, c. 1446 BC, *Everlasting-to-Everlasting,*
> *A Biblical Creation Account:* Psalms 90:1-10

And in the 86 AD, *Eternity-to-Eternity* Account by Peter:

The future mass extinction ending Eon 4:
[12a]   *Looking for and hasting unto the coming of the day of God,*
[12b] *wherein the heavens being on fire shall be dissolved,*
[12c] *and the elements shall melt with fervent heat?*

The Coming Judgment and destruction
[7]   *But the heavens and the earth, which are now, by the same intelligence (logic) are kept in store, reserved unto fire against the day of judgment and perdition of ungodly men.*

Human Existence in Eternity Future (New Heaven and New Earth)
[13a] *Nevertheless we, according to his promise,*
[13b] *look for new heavens and a new earth,*
[13c] *wherein dwelleth righteousness.*
[14a] *Wherefore, beloved, seeing that ye look for such things,*
[14b] *be diligent that ye may be found of him in peace, without spot, and blameless.*

God Existence In Eternity Future

    [15a] *And account [that] the longsuffering of our Lord [is] salvation;...*

    [18b] *Jesus Christ. To him [be] glory both now and to the day of eternity. Amen.*

Peter, c. 64-68 AD, *Eternity-to-Eternity,*
A Biblical Creation Account: 2Peter 3:1-18

And again in the 96 AD Time and Eternity Creation Account by John:

The future mass extinction ending Eon 4:

[10:6] *And sware by him that liveth for ever and ever, who created heaven, and the things that therein are, and the earth, and the things that therein are, and the sea, and the things which are therein, that there should be time no longer:*

[10:7b] *the mystery of God should be finished,*

The Coming Judgment and destruction

[14:7] *Saying with a loud voice, Fear God, and give glory to him; for the hour of his judgment is come: and worship him that made heaven, and earth, and the sea, and the fountains of waters.*

Human Existence in Eternity Future (New Heaven and New Earth)

[21:1] *And I saw a new heaven and a new earth: for the first heaven and the first earth were passed away; and there was no more sea.*

[21:2] *And I John saw the holy city, new Jerusalem, coming down from God out of heaven, prepared as a bride adorned for her husband.*

[21:3] *And I heard a great voice out of heaven saying, Behold, the tabernacle of God [is] with men, and he will dwell with them, and they shall be his people, and God himself shall be with them, [and be] their God.*

[21:4] *And God shall wipe away all tears from their eyes; and there shall be no more death, neither sorrow, nor crying, neither shall there be any more pain: for the former things are passed away.*

God Existence In Eternity Future

21:5 And he that sat upon the throne said, Behold, I make all things new. And he said unto me, Write: for these words are true and faithful.

21:6 And he said unto me, It is done. I am Alpha and Omega, the beginning and the end. I will give unto him that is athirst of the fountain of the water of life freely.

21:7 He that overcometh shall inherit all things; and I will be his God, and he shall be my son.

22:13 I am Alpha and Omega, the beginning and the end, the first and the last.

John, c. 90-96 AD, *Time and Eternity*,
A Biblical Creation Account: Revelation

The above quotes are from just some of the major creation accounts. It may seem ironic that creation accounts should include history of the future. However, in the context of the concept of God being everlasting and eternal, the creation itself includes all the eons of time, not just the beginnings of the past.

The above examples illustrate the harmony among all the creation accounts. These points of detail are not limited to just these accounts, but are the theme of the bible. It is what we are living for, our hope for the future. When our terrestrial bodies expire, our extraterrestrial existence will continue. We will outlive the physical creation itself.

Summary of Chronology and Detail.

Thus it is demonstrated in the complete agreement and harmony of all the creation accounts of the bible combined. When you interpret the *Seven-Day Creation Account* in the light of the rest of the bible, you get a completely different picture than when you interpret the rest of the bible in the light of the traditional interpretation of the *Seven-Day Creation Account*.

Not only do all the creation accounts of the bible combined present an internally consistent picture of technical detail and chronology, it is also in full agreement with the facts as discovered by modern science. There is no disagreement between the true facts as presented in the bible and the true facts as discovered by modern science. Any perceived disagreement is between the interpretations of the theologians, going beyond what the bible says, and the interpretations of the scientists, going beyond observed facts as discovered by science.

There are two major differences between the bible and science.

1.   The bible published the facts thousands of years before discovery and publication by modern science.

2.   The bible attributes it all to an intelligent creator God where modern science is of the opinion that it all arose spontaneously.

Appendix A: Detailed Chronology List

Adapted from *Eyewitness to the Origins, Third Edition*, 2009, Ch. 4:

Chronology and Detail According to the Bible

There are about three dozen major accounts of creation found in the bible. When all the chronology clues from all the creation accounts are combined the results present a clear order of events. That order of events is presented here

A clear biblical order of creation events becomes obvious when all the major biblical accounts of creation are combined. There are about three-dozen such accounts. When all the chronology clues from all these accounts are compared the final result is the same order of events as observed in reality. That order of events is presented here.

(Note: This is not to be confused with the six-day order of creation events as in the traditional, or "official" interpretation by theologians based on only a single one of those three-dozen biblical accounts, while ignoring all the others.)

This chronology of events and its technical details were unknown to humanity other than through the bible until their independent discovery by modern science. Yet, humanity was blinded to it.

In the meantime, the imagination and reasoning of humanity produced some absurd postulations concerning the beginnings. None of those human productions came close to the truth as actually recorded in the bible and later discovered by modern science.

Most of the independent discovery of truth came within the past three hundred seventy years. Some details are not yet widely accepted by the scientific community, some details may even be yet undiscovered.

This section compares the early publication in the ancient scripture to the later, more recent confirmation by modern science discovery. This section tries to avoid discussion of the erroneous traditional interpretations that divide truth seekers. Those are discussed in the history section of the book, Eyewitness To The Origins, available on the internet at www.EyeWitnessToTheOrigins.com .

Note: The Old Scientist does not claim this list to be exhaustive. He has gone over it several times and each time he has found more detail. This is an early edition. In the interest of expediency and the need to publish in a timely manner, this list and the numbering should be considered to be temporary.

Also Note: The Old Scientist does not claim infallibility. The Old Scientist would rather reserve the right to make a mistake and not place the validity of the rest of his work at risk. All statements must stand on their own, and not depend on the authority of The Old Scientist for their validity. Neither should the acceptance of validity rest on the credentials of any other human authority. If a statement is true, it is true. If a statement is false, it is false. No credentials of any authority, either human or divine, can make a falsity true. Therefore, a multitude of citations, common to most scientific works, is avoided in this work. The serious students can research and discover the simple truth for themselves, unbiased and un-intimidated by any credentialed authority. Any observations or suggestions should be forwarded to: SimpleTruth@AnOldScientist.com.

Organization of Chronology

In contrast to the traditional divisions of history, the technical details of the bible are more readily presented in a slightly different organization as suggested by the following quote. In that quote, the **"insemination of the cosmos"** is the event known to modern science as the "big bang." This event begins the physical existence of the universe. The **"consummation of the eon"** is what is generally translated to be the "end of the world."

> *"For then must he often have suffered since **the insemination of the cosmos**: but now once in **the consummation of the eon** hath he appeared to put away sin by the sacrifice of himself. And as it is appointed unto men once to die, but after this the judgment:"* [bold emphasis added]
> Anonymous, ca. 60–70 AD, *The Bible,* Hebrews 9:26, 27

These two events bracket the time of the physical existence of the universe within eternity.

To make the organization include all existence, assuming existence in time beyond the physical universe, the eons of eternity past and future are added to each end.

That gives us:

Eon of Eternity Past (Timeless Past)
Physical Existence of the Universe (Three Eons of Time)
Eon of Eternity Future (Timeless Future-Includes "The Judgment")

Both the creation accounts of the bible, and modern science recognize two significant events within the chronology of the physical existence of the universe.

The first of those two events is **the emergence of the continents**. It is a big deal in the bible. That same point in time is also a big deal in the division into eons according to modern science. That point in time closely relates to the beginning of the **geologic column**. Could this be an endorsement of the geologic column by the creation accounts of the bible? More likely, the discovery of the geologic column is an endorsement of the creation accounts of the bible. In either case, the bible made a big deal of it first.

The second of those two events is the **explosion of a multitude of complex life forms** on this planet earth. It also is a big deal in both the bible accounts and in modern science. It is the beginning of the Eon of Complex Life Forms. Modern science recognizes it and has named it the **Cambrian Explosion**.

These two significant events are dividing points that give us three eons within the time of existence of the current physical universe.

Adding the before and after eons of timeless eternity gives us five major Eons:

> Eon of Eternity Past – Timeless Past.
> Eon of Early Development – Begins with the "insemination of the cosmos".
> Eon of Preparation for Complex Life /Ecology – Begins with the emergence of the continents.
> Eon of Complex Life Forms – Begins with Cambrian Explosion, Ends with the "consummation of the Eon."
> Eon of Eternity Future – Timeless Future, including "The Judgment"

Within the technical details of existence there are many events and subdivisions of time found in the biblical accounts. These are presented, along with many details within each of those subdivisions, in chronological order according to the combined biblical accounts.

Contrary to the traditional interpretation, nearly all the time within these three eons of time is before the existence of humanity on this planet. According to the bible, humans did not appear on the scene until the last small fraction of the last of these three great eons. Yet, the bible demonstrates knowledge of extreme detail of technical fact and chronology of events that occurred during the vast expanse of time in those eons when humanity did not exist.

The Eon of Early Development spans from the darkness before the universe began and produces a universe complete with a planet fully equipped to be a habitat for life.

In the Eon of Preparation for Complex Life the cycles of nature essential to ecology were set in motion. In this eon life began to inhabit the planet earth.

In the Eon of Complex Life Forms large sea life appeared. The age of the dinosaurs—the pinnacle of habitability of the planet earth—came and went. Finally, near the end of that eon, the age of mammals prepared the ecology for human habitation. We are currently living in the later stages of that last eon.

Organization of Detail

For the chronology and points of detail within the chronology, a numbering system has been employed. Each detail is assigned a number containing several decimal points.

The numbering system indicates two things about each detail: 1. The source, bible or science, or both, and 2. Chronology, or lack thereof.

The source of the detail is indicated by the typeface:

1.1.1..1.1 – normal (roman) typeface, plus bold, indicates bible verified by modern science.
1.1.1..1.1 – Italics typeface, not bold, indicates bible only.
1.1.1..1.1 – normal (roman) typeface, not bold, indicates modern science only.

The chronology or lack thereof of each detail is indicated by the decimal locations:

The chronology is indicated by up to three digits separated by single decimal points.

The first digit is the Chronological Eon.
The second digit is the Era, Age or Major Event within that Eon.
The third digit is the chronology within that age or event.

The lack of relative chronology among multiple details within a subdivision of time is indicated by numbering after a double decimal. An example of such simultaneous or non-chronology detail would be during Eon 2, Age 4, Event 1, the development of the molten magma outer layer of the planet earth. During that event, the following details mentioned in the bible and denoted by a number after the double decimal, were present in no significant chronological order.

The following is an example illustrating the format:

2.0.   Eon of Early Development

2.1.   **The Universe Formed -** *The Heavens*
2.2.   **The Expansion of the Universe**
2.3.   **The Development of the Solar System**
2.4.   **The Development of the Planet Earth**
2.4.1..   Then planet earth developed with molten magma surface (mantle.)
2.4.1..1   Mantle Planned to support the earth (continents.)
2.4.1..2   Mantle Engineered to support the earth (continents)
2.4.1..3   Molten mantle ("molten support" (pillars)) develops, to support the earth (continents)
2.4.1..4   Solidified Mantle is destined to become the ocean floor
2.4.1..5   Held in place by gravity/density (floatation)
2.4.1..6   Without continents
2.4.1..7   And without soil
2.4.1..8   Without oceans.
2.4.1..9   Without surface water
2.4.1..10   No humans existed
2.4.1..11.   With Juvenile water coming from inside the earth
2.4.2..   Then after a while the earth cooled enough to develop surface water

Following the double decimal after the chronology indicator may be a detail number that is not indicative of chronology, sometimes also followed by a sub-detail number after another single decimal. That sub-detail is also is not indicative of chronology, but is related to something within that leading detail within that event or age.

Chronology and Detail from All Biblical Accounts:

1.0. Eon of Eternity Past

1.0.0   Eternity Past Existed

1.0.0..1 Timeless [endless] Eternity Past Existed

1.0.0..2 Before the Universe existed

1.0.0..3 Something (God) existed in the absence of the universe [Intelligence, Wisdom, and Power]

1.1.0   Conditions Were Set Up to Start Universe

1.1.0..1 That Something (God) caused the Universe to form.

1.1.0..1.1 Planned by intelligence [Scientist aspect of pre-existence: Laws of Physics, Logic]

1.1.0..1.2 Designed by Wisdom [Engineer aspect of pre-existence]

1.1.0..1.3 Constructed (created) by Power [Authority (ruler, king, lord, father) aspect of pre-existence]

2.0. Eon of Early Development

2.1.   The Universe Formed - *The Heavens*

2.1.0..   The beginning of Universe was a specific event, the beginning of the current universe of time, space, matter and energy.

2.1.0..1.   The Universe had a beginning. It has not existed forever.

2.1.0..2.   It happened a long time ago

2.1.0..3.   Light was one of the first things to come into existence (Light existed from the "beginning")

2.1.0..3.1.   Light had a beginning. It has not shone forever.

2.1.0..3.2.   Light had a beginning from "darkness"

2.1.0..4.   It was an explosive event.

2.1.0..4.1   The bible describes it as a great blowout.

2.1.0..4.2   Modern science refers to it as a "Big Bang."

2.2.   The Expansion of the Universe

2.2.1..   Then there was the expansion of the universe.

2.2.2..   Then water became abundant in outer space

2.2.3..   Then stars came into existence

2.2.4..   Then the Sun came into existence

2.3.   The Development of the Solar System - *Chambers of the South*

2.3.1..   Then the solar system developed

2.3.1..1.   The earth and other planets developed.

2.3.1..2.   The planets (chambers) store water to moderate the cycle

2.3.1..3.   Water had accumulated in solar system

2.3.1..4.   Before life on earth

2.3.2..   Then hydrologic cycle (water cycle) developed in outer space of the solar system.

2.3.2..1.   That outer space hydrologic system is permanent

161

2.3.2..2. Intelligently designed for a purpose, To fill and maintain future oceans of planet earth.

2.3.2..3. Water carried in by small comets

2.3.2..4. Solar winds existed in solar system

2.3.2..4.1. Water carried out by solar winds.

2.3.2..4.2. Visible glow is a characteristic of solar winds

2.4. The Development of the Planet Earth

2.4.1.. Then planet earth developed with molten magma surface (mantle.)

2.4.1..1 Mantle Planned to support the earth (continents.)

2.4.1..2 Mantle Engineered to support the earth (continents)

2.4.1..3 Molten mantle develops, to support the earth (continents) (the "molten support" (pillars))

2.4.1..4 Solidified Mantle is to become the ocean floor

2.4.1..5 Held in place by gravity/density (floatation)

2.4.1..6 Without continents

2.4.1..7 And without soil

2.4.1..8 Without oceans.

2.4.1..9 Without surface water

2.4.1..10 No humans existed

2.4.1..11. With Juvenile water coming from inside the earth

2.4.2... Then after a while the earth cooled enough to develop surface water

2.5. The Development of the Atmosphere

2.5.1.. Then an atmosphere developed.

2.5.1..1. Atmosphere captures water from outer space.

2.5.1..2. Watered the surface of planet earth

2.6. The Development of the Oceans - *The Deep*

2.6.1.. Then the atmosphere captured (separated, set apart) water from outer space.

2.6.1..1 Adding to original juvenile water

2.6.1..2 Filling the oceans

2.6.1..3 Separating (extracting) water below the atmosphere (oceans)

2.6.1..4 (direction of separation (extraction) was from above to below)

2.6.1..5 from water above the atmosphere (small comets theory.)

2.6.2.. Then the earth was covered by the oceans,

2.6.2..1. Solidified mantle became ocean floor

2.6.2..2. Intelligence existed prior to this

2.6.2..3. Engineering existed prior to this

2.6.2..4. No continents existed at this time.

3.0. Eon of Preparation For Complex Life / Ecology - *The World*

3.1. The Emergence of the Continents - *The Earth (Land)*

3.1.1..   Then catastrophic event(s) occurred (Mantle overturn event(s)?)
3.1.1..1.   Sea level established relative to continents (Sea held back)
3.1.1..2.   Continents (land) emerged ("brake forth," "issued out," etc…)
3.1.1..3.   Storm-Clouds-Wind
3.1.1..4.   Darkness (debris in atmosphere)
3.1.1..5.   Tectonic Activity: Rapid subduction (breaking of tectonic plate?)
3.1.2..   Then Continents began to emerge
3.1.3..   Then Continents established to be above sea level
3.1.4..   Then Continents surrounded by continuous ocean,
                    not oceans surrounded by continuous continent.
3.1.5..   Then As continents grew there were episodes of mountain building,
3.1.5..1.   Mountains rose supported by floating on molten mantle
                    (Isostatic Rebound)
3.1.5..2.   Valleys fell
3.1.5..3.   Isostatic balance developed
3.2.   **Cycles of Nature Established as the Basis of Enduring Ecology**
3.2.0..   An era of equilibrium after continent and mountain building
3.2.1..   Lithologic Cycle Established - *To Maintain Continents*
3.2.1..1.   Continent Maintaining Mechanism becomes Established
3.2.1..2.   Soil profile formed
3.2.2..   Hydrologic Cycle – *Waters the Continents*
3.2.2..1.   According to pre-established laws of physics
3.2.2..2.   Hydrologic Cycle for purpose of sustaining life
3.2.2..3.   Sustained by water from the chambers. (incoming cosmic water)
3.2.2..4.   Hydrologic Cycle maintained for Future Time
3.2.2..5.   Earth ready for life to begin,
3.2.3..   Then Life begins on earth – green things start growing
3.2.4..   Food chain—Carbon Dioxide/Hydrocarbon Cycle develops
3.2.4..1.   Photosynthesis Develops for CO_2 /Hydrocarbon cycle
3.2.4..2.   Photosynthesis becomes the basis of the food chain
3.2.4..3.   The atmosphere became habitable by photosynthesis
3.2.4..4.   Food Chain becomes fully developed
3.2.4..4.1   Ultimate purpose of food chain is for sustaining human life.
3.2.5..   Reproduction Cycle—Biogenesis
3.2.5..1.   Life became abundant on earth
3.2.5..1.1.   Plant biogenesis developed
3.2.5..1.2.   Animal biogenesis developed
3.2.6..   Chronobiology Cycles—Chronobiology Rhythms
3.2.6..1.   Lunar Cycle for Seasons
3.2.6..2.   Circadian/Annual Cycles
                    (Day/Night/Seasonal Patterns of Life Forms)

3.2.7..   Food chain (Carbon Dioxide/Hydrocarbon cycle) fully developed

3.2.8..   Mass Extinction/Sustained Ecology Cycle—
                     Punctuated Equilibrium

3.2.8..1.   Ecology Destruction and Renewal Became Pattern of Existence

3.2.8..1.1.   with mass extinction

3.2.8..1.2   by inundation

4.0.   Eon of Complex Life Forms

4.1.   **Peak of Habitability—Ecology of Global Warming**

4.1.1..   Era of sustained ecology.

4.1.1..1.   The sea swarms with living creatures.

4.1.2..   Then Plants with seeds developed.

4.1.3..   Era of Megafauna (Monsters)

4.1.3..1.   Megafauna (dinosaurs) thrive

4.1.3..2.   Pinnacle of Nature/Creation

4.1.3..3.   Sea monsters (Leviathan) thrive

4.1.3..4.   First birds developed

4.1.3..5.   First Flowering (Seeding) Plants.

4.1.3..6.   Ecology continues

4.2.   **Major Mass Extinction (Cretaceous/Cenozoic Boundary Event)**

4.2.1..   Major Mass extinction (Mantle turnover event with extinction)

4.2.1..1.   Tectonic Activity End Cretaceous

4.2.1..2.   Catastrophic event (great storm) (tidal waves)

4.2.1..3.   Darkness (debris in atmosphere)

4.2.1..4.   Breakup of continents (Mantle overturn event?)

4.2.1..5.   Great flood *(breath taken away)*

4.2.1..6.   Extinction (death/dust)

4.2.1..7.   Isostatic rebound (mountain building episode)

4.3.   **Ecology Restored, less Global Warming, Rise of Mammals**

4.3.1..   Restoration of ecology - *Renewest the face of the earth*

4.3.1..1.   New life forms Cycle Of Extinction & Renewal
                     (Punctuated Equilibrium).

4.3.1..2.   Spread over continents The far past of recent life.

4.3.2..   Era of sustained (renewed) ecology.

4.3.2..1.   Mammals appear

4.3.2..2.   Forests and grasslands

4.3.2..3.   Human species appears on earth

4.3.2..4.   Man became a living soul

4.3.2..5.   Tropical ecology with mild global warming

4.3.2..6.   Punctuated Equilibrium Continues

4.4.   **Another Mass Extinction (Miocene/Pliocene Boundary Event)**

4.4.1..   End of Tropical Ecology - *End of Garden of Eden*

4.4.1..1. Destruction of Habitat/Ecology event

4.4.1..2. Tectonic motion event

4.4.1..3. Volcanism event

4.5.   Ecology Restored, Global Cooling

4.5.1.. Era of restored ecology (Long time passes.)

4.5.1..1. Climate turns colder and harsher - Global Cooling

4.5.1..2. Thorns and Thistles ecology (carbon starved ecology)

4.5.1..3. Expansion of humanity after habitat/ecology destruction -
Replenish

4.5.1..4. Dedication of food chain

4.5.1..5. Photosynthesis is the basis of the food chain for all forms of life.

4.5.1..6. Pre-historic Past of Humanity

4.5.1..7. Rate of Extinction exceeds Rate of Speciation
(Creation Complete)

4.5.1..8. Long Time Passes. Pleistocene

4.5.1..9. Rise of language and civilization

4.6.   Another Mass Extinction
(Pleistocene/Holocene Boundary Event)

4.6.1..1. Mass Extinction of Large Mammals

4.6.1..2. Mass Extinction was in form of a flood (*Noah's*)

4.7.   Ecology Restored/Epoch of Secular Human History

4.8.   Era of restored ecology
Proliferation of civilization after near extinction

4.8.1.. Beginning times/Current Era Boundary

4.8.2.. Living in Present

4.8.2..1. "End Times" Philosophy (false)

4.8.2..2. Uniformitarianism Philosophy (false)

4.8.2..3. Logic of God, Laws of Physics (true)

4.8.2..4. Present time/Future time boundary

4.8.3.. Living In Future

4.8.3..1. Time of Sustained Ecology.

4.9.   End of Era of Complex Life Forms Event

4.9.1..1 There will be an end.

4.9.1..2 A Future Mass Extinction – Great and Terrible Day of the Lord.

5.0. Eon of Eternity Future

5.0.0.1. Judgment

5.0.0.2. New Heaven & Earth.

5.0.0.3. Eternity Future Will Exist.

5.0.0.4. Human Existence In Eternity Future.

5.0.0.5. God Existence In Eternity Future.

Appendix B: Summary Chronology List

Chronology: From Eternity Past to Eternity Future
Eon of Eternity Past (Timeless Past)
Something Existed before the Universe came into existence
Conditions were set up for the universe to come into existence
Eon of Early Development
The beginning of the universe event, "Big Bang," "Shout," etc.
The expansion of the universe
The development of the solar system
The early development of the planet earth
The planet earth with hot molten surface
The planet earth without any surface water
The first juvenile surface water springing (out gassing) from
    volcanoes
The development of the atmosphere
The development of the oceans
The ocean covering the surface of the planet earth
Eon of Preparation for Complex Life/Ecology
The emergence of the continents
Cycles of Nature Established as the Basis of Enduring Ecology
Lithologic Cycle
Hydrologic Cycle
Carbon Dioxide/Hydrocarbon Cycle (Food Chain)
Cycle of Life (Biogenesis)
Circadian Cycle
Cycle of Mass Extinction/Sustained Ecology (Punctuated
    Equilibrium Cycle)
Eon of Complex Life Forms
Era of Sustained Ecology – Pinnacle of Ecology, Global Warming,
    Megafauna
Major Mass Extinction Event (Correlates with Kt Boundary event)
Ecology restored – Rise of mammals, less global warming
Sustained Ecology Period – Garden of Eden, Global tropical
    ecology
Minor Mass Extinction Event – End of Garden of Eden, End of
    global tropical ecology
                (Correlates with Miocene/Pliocene Boundary Event)

Sustained Ecology Period – Thorns and Thistles Ecology, Global Cooling

Another Mass Extinction Event – (Noah's Flood?)
(Correlates with Pleistocene/Holocene Boundary Event)

Sustained Ecology Epoch (Time of Secular Human History Present and Future)

End of Era of Complex Life Forms Event – Future Mass Extinction

Eon of Eternity Future (Timeless Future)

New heaven and new earth

Appendix C: Does Nothingness Exist?

What really existed before the universe began? Lets start at the beginning, or as some might say, before the beginning. What existed before the universe began? What about nothingness[73]

Does nothingness really exist? That is a problem that has bothered me since early childhood. I remember many years ago, At least sixty-five years ago, because I had to be less than ten years old, I remember looking out at the night sky and asking my older sister, "What is out beyond the stars?" Of course, she thought she knew the answer. "The universe is a great circle, if you go out far enough, you end up back where you started." Well, I was not satisfied with that answer. Beyond the universe, there had to be either something, or nothing.

But, does nothingness really exist?

Recently, I was watching a video of a presentation where the speaker was explaining the beginning of the universe. He said something like, "Nothingness, nothingness, nothingness, Poof, everything." That was his explanation of what scientists have called the "big bang" or theologians have called the creation "Ex Nihilo."

So, what about the claim by some theologians that God created the heavens and the earth "Ex Nihilo" (from nothing)?

In any event, that is not a very satisfying explanation either. That set me to dreaming about the existence of nothingness…

Scientists have speculated that something had to exist (in the "nothingness" mentioned above) to set up the conditions for the "big bang" to occur.

The bible says something existed before the universe was created. It says there was some living existence and named it with various names having the basic meaning of, that which exists forever, or that which possesses three distinct characteristics, the first characteristic being of influence, force, energy or power, the second being logic or intelligence, and the third, being the prudent use of knowledge, or, as some would call it, "wisdom."

---

[73] *Excerpt from article,* Does Nothing Exist? *Science and the Bible Ezine Volume 3, Number 5, May, 2010, http://www.scienceandthebible.net/ezine/2010/SATB-Ezine-2010-05-31.htm*

So, it seems that both science and the bible agree on one thing. Something had to exist in the nothingness before the beginning of the universe as we know it today.

Uh, Oh. We are getting dreadfully close to saying that nothingness did not exist in the archaic past before the existence of the universe. That might upset some theologians. Where (or when) they would assume nothingness to exist, there was at the least, influence, force, energy, or power of some kind. And influence, force, energy, or power, is not nothingness.

But what about in the realm of space? I have heard it said that inside the molecule, in a single atom, the actual atomic parts of the atom occupy only a tiny fraction of the space. The rest is mostly nothingness. However, I ask, How can there be nothingness there when the parts somehow are affected by the influence of all the other parts in such a way that they do not go on their merry way, abandoning the other parts? Don't they seem to be held together by some kind of influence, force, or energy? If the parts of the atom cannot escape, that influence, maybe there is something in that vast nothingness within the infinitesimal atom that cannot be escaped, something that is omnipresent, eliminating the existence of nothingness within the atom, even if it is only some kind of influence.

On the larger scale, but still small beyond imagination, what about the nothingness of space outside each atom and between the atoms and molecules that make up the stuff we can see, touch, and smell? There exists in that supposed nothingness, the influence that holds those atoms and molecules into the matter that makes up the world we all know and enjoy. Without that influence holding the atoms of our automobile together, how could we enjoy going for a Sunday drive?

But then, nothingness must exist in the vastness of outer space of the universe, beyond the atmosphere of our planet earth, and before we get to the sun or other planets??? Yet, that vast emptiness is filled with the influence of gravitational attraction. If there were some place in the vastness of empty space of our solar system, where that influence did not exist, our planet might fall into it and go flying out of the solar system. No, there is nowhere in the universe that nothing exists. The universe contains the omnipresence of some form of energy, force, or influence that holds it together.

But the universe is expanding, flying apart at breakneck speed. Could it be that beyond the limits of the current universe of matter and energy, there is space occupied with nothingness? If there was a time, before the beginning of the universe, that nothingness existed, with infinite dimensions, (very big nothingness) and if the influence of energy is limited by the speed of light, could it still be possible that nothingness exists beyond the limits of the current universe, beyond where the influence of energy could have traveled at the speed of light since the beginning of the universe?

Well, I never claimed to know everything.

And our dog woke us up before the end of the dream.

But, scientists say that the first law of thermodynamics says energy can be neither created nor destroyed, that is to say, energy is eternal. And they also say that something had to exist before the beginning of the universe as we observe it today, to set up the conditions for it to come into existence. And they also say that mater (the stuff that makes up the physical universe, the elements and molecules etc.) and energy, are somewhat interchangeable, that is, they can switch from matter to energy, or back, through the equation, $E=MC^2$. So, if the stuff that makes up the universe existed before the universe, and the universe is fully occupied with stuff such as gravity, and at least some of that stuff cannot be either created or destroyed, that means there never has been real estate for nothingness to exist, at least in this universe or its predecessor.

So, what about the claim by some theologians that God created the heavens and the earth "Ex Nihilo" (from nothing)?

But, the bible says that God is eternal, that is God exists infinitely in time, long before the universe, and forever into the future, and theologians claim that God is omnipresent, that is God is everywhere at the same time. So, that means that there is no place for nothingness to exist.

Unless, of course, theologians are saying that God is nothing.

But that does not coincide with what the bible has to say on the subject.

And modern science agrees with the bible that something had to exist to set up the conditions for the universe to come into existence.

Appendix D: Words Have Meanings

Technical Meanings of Religious Words and Phrases

By Max B. Frederick, AnOldScientist

A Glossary of Technical Meanings of Religious Words and Phrases is found on the internet at, www.ScienceAndTheBible.net/glossary.

The first rule of biblical interpretation: **Words have meaning**.

If you want to know what the bible has to say in the realm of religion, you first have to know the religious meanings of the words.

If you want to know what the bible has to say on any subject that is in the realm of science, you have to figure out what the words mean in the technical realm rather than the religious realm.

A persistent problem is the assumption that the bible is about religion, not about reality. Reality is assumed to be in the realm of science.

There is a saying that is popularly being passed around that says something like, "The bible is not a science book, but where it touches on science it is right." I disagree with the way that statement is improperly used to dismiss the subject of science and go on to ignore the questioning of intellectually honest doubters.

The topic of the ultimate origins—the creation-evolution debate—is where the bible not only touches on a subject that is in the realm of science, It delves into it deeply.

The bible has a lot more technical stuff to say on that topic than most theologians ever dreamed. In the bible there are nearly three-dozen major accounts of creation, and many minor ones.

Yet, the consensus[74] among theologians as to what it says on that subject is so far from what it really says that their traditional interpretation has become known as the Judeo-Christian Creation Myth and intellectually honest scholars have relegated it to the status

---

[74] This religious method of deciding reality by the agreement of respected authorities is also rampant in science and is known as, Science by Consensus, and is not in accord with the scientific method of true science.

171

of myth along with the creation myths of other religions. Then the argument arises: If you cannot believe the first thing the bible says, how can you believe the rest of the bible? That has been a major stumbling block to intelligently honest scholars, including our children who have been brought up in the church and then go to college and learn what they have been taught the bible says is not true.

The problem is not in what the bible really says.

The problem is, most of what the bible really says is ignored. It is ignored for several reasons. In no particular order those reasons are:

The people doing the biblical interpretation are theologians, not scientists and they do not recognize scientific principles when they see them—they do not even know the technical meanings of the words in the original language.

Theologians are inclined to believe that there is only one account of the origins in the bible. In fact, there are many such accounts— almost three dozen of them are major accounts, backed up by many more minor mentions of it.

Most of those accounts are ignored because those biblical accounts do not agree with the pre-conceived idea of what should be said on that topic. And there is where the pre-conception includes what is known as the Judeo-Christian Creation Myth rather than what the rest of the bible actually has to say on the subject.

Theologians have adopted the politically correct stance that the traditional interpretation of the one single account that is up front in the bible is the "official" biblical schedule of creation, even though there are many major biblical accounts that disagree with that interpretation.

Many of the things recorded in the ancient scriptures of the bible have historically been unbelievable because the reality of it had not yet been discovered by modern science. In many cases it can be demonstrated that since they could not believe what it really says, they simply fudge the translation (interpret it) to agree with reality as they saw it at the time of translation—or at the time they practiced science by consensus as in the trial of Galileo for which they had to issue a formal apology some three hundred fifty years later.

But the big reason is, words really do have meaning, and many of the words mean something unfamiliar to theologians. In the original language they have technical meanings that have been ignored. Those technical meanings have been unrecognized, or they have been simply ignored, in favor of religious meanings.

If you want to understand what the bible has to say about things in the realm of science, to be sure of what it says you must apply the scientific method rather than the religious method. The scientific method is to state what it might say, then test that hypothesis against reality. The religious method is to say what you think it might say then get the opinion of someone with more credentials than yourself. Science by consensus is the religious method.[75]

Reviving an ancient language that has gone through long periods of not being used is an art. The meanings of many words have been lost. The good part of it is, when a language is not being used, the meanings of words do not change.

A Glossary/Lexicon of Technical Meanings of Religious Words and Phrases is found on the Internet at

www.ScienceAndTheBible.net/glossary.

That Glossary/Lexicon has been a long time in the development, and still is in that process of development. That is why it is not in this book, but on the Internet so it can be frequently updated. It is the product of thousands of hours of studying science and the bible. Some of the entries are from the third edition of the book, *Eyewitness to the Origins*, published in 2008, after over fifteen thousand hours of study, the equivalent of about seven and a half years of full time work. And that was over six years ago. Many of the entries were published in the periodical, *Science and the Bible Ezine* during the past six years. Those are the result of recognizing the discrepancy between the religious interpretation of the ancient scriptures and the reality found therein. Comparing the reality discovered by modern science and how those words were actually used in those ancient writings of the bible sheds new light on the meanings of many

---

[75] *There is more to it than that and a treatise on the two methods can be found in the book, Eyewitness to the Origins…*

words—meanings that had been lost to, or obscured by, religious interpretation.[76]

When you interpret the whole bible in the light of the assumption that the six days of Genesis 1 you get a whole different bible than when you interpret the six days of Genesis 1 in the light of what the whole bible really says about the origins. To get it right you must understand the technical meanings of the words.

---

[76] *See Ezine January 2010 about how ancient information traveled through time to arrive where it is today.*
http://www.scienceandthebible.net/ezine/2010/SATB-Ezine-2010-01-31.pdf

www.ingramcontent.com/pod-product-compliance
Lightning Source LLC
Chambersburg PA
CBHW071430180526
45170CB00001B/281